SpringerBriefs in Astronomy

Series Editors

Martin Ratcliffe
Wolfgang Hillebrandt

For further volumes:
http://www.springer.com/series/10090

Helmut Lammer

Origin and Evolution of Planetary Atmospheres

Implications for Habitability

 Springer

Helmut Lammer
Space Research Institute
Austrian Academy of Sciences
Graz
Austria

ISSN 2191-9100   ISSN 2191-9119   (electronic)
ISBN 978-3-642-32086-6   ISBN 978-3-642-32087-3   (eBook)
DOI 10.1007/978-3-642-32087-3
Springer Heidelberg New York Dordrecht London

Library of Congress Control Number: 2012945127

Printed on acid-free paper

Springer is part of Springer Science+Business Media (www.springer.com)

Foreword

Advances in exploration of the atmospheres of the Solar System planets achieved during recent decades by means of numerous successful American, Soviet, European and Japanese space missions, as well as recent spectacular discoveries of hundreds of new extrasolar planets orbiting distant stars, have worked as a powerful impulse for promoting further studies of the formation and evolution of the atmospheres of the Solar System planets and extrasolar planets. Theoretical analysis of space observations of the planetary systems of stars having different ages and different spectral classes is also an essential part of the studies of conditions favorable for the origin of biospheres similar to that on Earth, that is, of other habitable worlds. Such analysis includes, above all, studies of the impact of the Sun on the upper atmospheres of the Earth and of the other Solar System planets. This impact takes place due to the absorption of the solar short-wave electromagnetic radiation, as well as to the solar corpuscular fluxes. So, to understand and explain the effects of the solar radiation on the upper atmosphere, a detailed study is warranted of the interaction of energetic short wave solar photons with atmospheric atoms and molecules that results in numerous ionization, dissociation, and excitation processes of both neutral and ionized atmospheric constituents. Also, studies are further needed of atmospheric chemical reactions, the interaction of the ionosphere plasma with the planetary magnetosphere and plasma fluxes ejected by the extended solar corona which are streaming past a planet, of thermal and non-thermal atmospheric loss processes, as well as of numerous other physical and chemical processes generated in the upper atmosphere by the Sun. All these processes and phenomena are a subject of a relatively new branch of atmospheric science which arose with the advent of the Space Era and which is called by its godfathers—celebrated geophysicists, Marcel Nicolet and Sydney Chapman—aeronomy. As contrasted to meteorology which studies the properties and behavior of the lower atmosphere between the surface and the tropopause where the weather phenomena are generated, aeronomy is a division of atmospheric science that studies physics and chemistry of the upper atmosphere that extends from above the troposphere up to the altitudes where it is modified by the solar wind plasma.

The subject of the present brief monograph, in full concord with its title, is connected with the studies of the physics and chemistry of the planetary protoatmosphere formation processes during and after the termination of a planetary accretion and with the studies of the following evolution of the protoatmosphere. The monograph begins with a discussion of the most likely scenarios for the formation of the primary atmospheres of terrestrial planets, that may be related to a collapse of the protosolar nebula, to catastrophic outgassing of the dense steam atmosphere from a solidifying magma ocean and the following condensation of a liquid water ocean, as well as to the volcanic outgassing of a secondary atmosphere of a planet as a result of its tectonic activity. The evolution of these original atmospheres due to a multitude of thermal and non-thermal atmosphere loss processes is then considered over geological time scales. The atmospheric escape is driven by powerful solar or stellar X-ray and extreme ultraviolet radiation, as well as by solar or stellar plasma coronal outflows felicitously called by Eugene Parker the solar or stellar winds. The evolution of the solar and stellar radiation and plasma environment of the host G-, K-, and M-type stars during their lifetime on the Zero-Age-Main-Sequence, which directly affects planetary atmospheres, their escape, and evolution, is also discussed.

The central part of the monograph presents a detailed discussion of the atmospheric loss mechanisms due to the action of various thermal and non-thermal escape processes for the neutral and ionized particles from a hot, extended atmospheric corona. Scenarios for the formation and evolution of the atmospheres of Earth, Venus, and Mars, that is, the planets orbiting within the habitable zone around the Sun, are considered. A crucial role of the magnetosphere of a planet in protecting its hot, extended, and partially ionized corona from the solar wind erosion is discussed. Here the important novel theoretical results presented are the results of simulation of atmospheric loss during the initial period of the evolution of the primary planetary atmosphere, as well as during the entire period of evolution—from the initial atmosphere formation until the present. The monograph has a number of new interesting ideas that may serve as starting points for further research. For example, one such idea is to use future observations of the UV transits of terrestrial exoplanets, orbiting M-type dwarf stars, from already planned space missions for verification of the existing and newly developed theories of the evolution of planetary atmospheres.

The book presents a brief review of the present state of knowledge of the aeronomy of planetary atmospheres and of their evolution during the lifetime of their host stars by taking into account conventionally accepted concepts, as well as recent observational and theoretical results. It also highlights tendencies for development in this rapidly evolving area of space science. The results of the study of aeronomy and evolution of planetary atmospheres, as well as a wide range of related problems presented and discussed in the monograph from a unified point of view, can be of interest for students and young scientists who would like to get some first knowledge of this area of atmospheric science and also for experienced researchers working in adjacent areas of planetary science.

The results of planetary research presented in the monograph were obtained by a large number of authors from different countries, mainly during the three last decades, although some important results were obtained by the science community considerably earlier. However, the majority of the results, which can be referred to in the bibliography, were obtained during the last decade. During this period Dr. Lammer published a large number of the results presented in the monograph, which were obtained either by him personally, or by an international team of collaborators from different countries in Europe, Russia, the United States, and Japan, where Dr. Lammer was usually coordinating the work. For example, the author of this foreword had the pleasure to cooperate with Dr. Lammer during recent years within a joint project supported by the Russian and the Austrian Academies of Sciences. Dr. Lammer also has provided a remarkable personal contribution to the studies of the atmospheric loss due to various thermal and non-thermal escape processes of neutral and ionized atmospheric particles from the hot, extended atmospheric coronae, which are discussed in his brief monograph.

Murmansk, June 2012   Yuri N. Kulikov

Preface

After more than four decades of space exploration and data gathering by various spacecraft, plausible hypotheses can be developed which may answer why Venus and Mars evolved so differently compared to the Earth. The Earth is unique in the sense that it has a nitrogen-dominated atmosphere, kept its liquid water oceans since ~ 4.4 Gyr ago, has continents and has been geophysically active since its origin. Due to these conditions, approximately 3.5–3.8 Gyr ago simple microbial and later complex multi-cellular life forms could originate and inhabit the planet's hydrosphere, subsurface and surface. Up to now, Earth is the only example of a known habitat where such a great variety of life forms could develop.

One important factor in this cosmic puzzle regarding Earth's evolution to a habitat where higher life forms could evolve is related to the fact that the planet orbits around the Sun, which is a GV2 star, within the so-called continuously habitable zone. The classical concept of the stellar habitable zone is a spherical shell around a main sequence star where a planet with an atmosphere can support liquid water at a given time. The width and location of this region depends on the stellar luminosity that evolves during the star's lifetime and is closer, compared to solar-like G-type stars, when the star is cooler (M- and K-types) and further out when the star is hotter (F-type).

Venus orbits slightly outside the inner edge of the habitable zone at a distance where the planet's initial water inventory most likely always remained in vapor form due to greenhouse conditions until it was lost to space. Mars on the other hand orbits at the outer edge of the habitable zone, where a maximum greenhouse effect fails to keep the global surface temperature of the planet above the freezing point of H_2O so that CO_2 can condense. Because Earth has also been geophysically active since its origin the carbonate–silicate cycle works, and CO_2 could be weathered out of the atmosphere via its continents and oceans into a carbonate deposit in the lithosphere. The amount of CO_2 in the atmosphere depends on the orbital distance where CO_2 can be considered as a trace gas close to the inner edge of the habitable zone but a major compound in the outer part of the habitable zone.

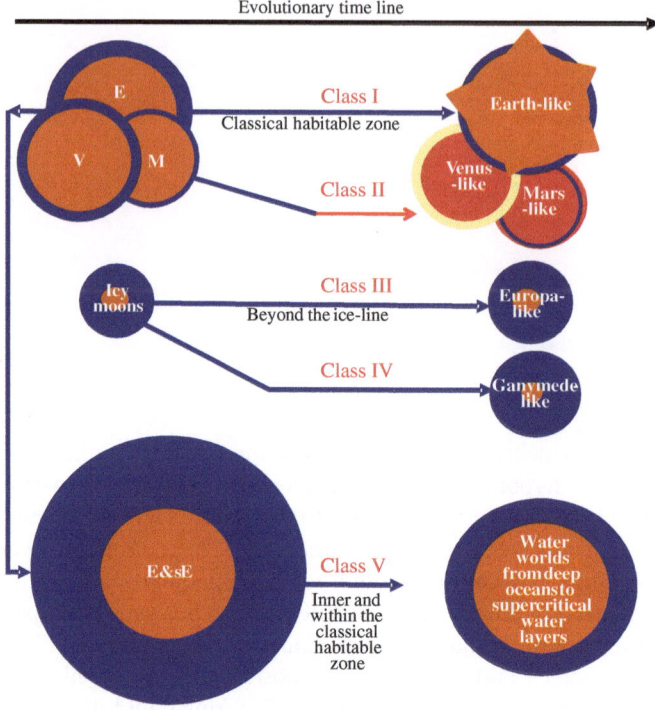

Fig. 1 Illustration of Earth-analogue class I, martian-type class II, icy moon-type classes III and IV and water world class V habitats

Due to my main scientific expertise in atmospheric escape processes and the evolution of planetary atmospheres, in 2003 I became involved in the European Space Agency (ESA) Terrestrial Exoplanet Advisory Team (TE-Sat). One of the main tasks within this international team was related to the extension of comparative planetology from Solar System planets to exoplanets, aeronomy and habitability. It soon became clear that the classical concept of the habitable zone and its related questions of what makes a planet habitable is much more complex than having a big rocky body located at the right distance from its host star. Because the classical habitable zone concept does not really indicate if life could evolve on a terrestrial planet, it is better to classify potential habitats (see Fig. 1). According to an interdisciplinary study which was coordinated by me and published in 2009 in *Astronomy and Astrophysics Review* (17:181–249, 2009) four potential habitat classes have been identified:

- Class I habitats represent planetary bodies on which stellar and geophysical conditions allow Earth-analogue planets to evolve so that complex multi-cellular life forms may originate and inhabit the planets hydrosphere, surface and subsurface environments.

- Class II habitats include bodies on which life may evolve but due to astro-physical and geophysical conditions, these planets rather evolve within their habitable zones towards Venus- or Mars-type worlds where complex multi-cellular life forms may not develop.
- Class III habitats are planetary bodies where subsurface water oceans which interact directly with a silicate-rich core exist below an ice layer.
- Class IV habitats have liquid water layers between two ice layers or liquids above ice.

However, since the publication of this study, exoplanets such as GJ 1214b or Kepler 11b have been discovered which may represent an additional fifth habitat class. From the radius–mass relation of these super-Earth-type planets one can expect that a rocky core is most likely covered by a very deep liquid water ocean but without continents. Thus, this new type of planets can be classified as:

- Class V habitats which correspond to bodies which have huge water layers above a rocky core but no solid surface. In some cases the environmental conditions may allow envelopes of supercritical water above a rocky and per-haps Earth-like nucleus.

A careful study of various astrophysical and geophysical aspects indicate that Earth-analogue class I habitats have to be located at the right distance of the habitable zone from their host stars, must lose their protoatmospheres during the right time period, should maintain plate tectonics over the planet's lifetime, should have nitrogen as the main atmospheric species after the stellar activity decreased to moderate values and finally, the planet's interior should have developed conditions that an intrinsic strong global magnetic field could evolve.

The recent discoveries of numerous planetary candidates by NASA's Kepler space observatory indicate that there may be millions of smaller terrestrial-type planets within orbit locations inside the habitable zones of their host stars in the Galaxy. However, preliminary but careful studies of potential habitats point in a direction that class I habitats should occur much less frequently compared to more exotic class II, III, IV and V habitats.

This brief monograph addresses the physical and chemical processes that underpin these findings for an interdisciplinary scientific readership and students which are interested in habitability and the escape-related evolution of planetary atmospheres. Initially, different hypotheses about the origins of protoatmospheres are discussed. Because the escape and evolution of atmospheres are strongly connected to the radiation and plasma environment of the age of a planet's host star the latest knowledge in the stellar age-activity relation is discussed. Therefore, the most important physical and chemical processes which are responsible for the evaporation and erosion of planetary atmospheres are described in detail. Atmo-spheric evolution scenarios for early Venus, Earth, Mars, terrestrial exoplanets and the implications for the search of Earth-analogue class I habitats are also addressed. Finally, powerful methods which are based on future UV transit observations of Earth-size exoplanets within orbits of dwarf stars together with advanced numerical modeling techniques, which can be used for the test of the atmosphere evolution hypotheses, are presented.

Graz, Austria, June 2012   Helmut Lammer

Acknowledgments

The author acknowledges support by the Austrian Research Foundation FWF/NFN project S116 "Pathways to Habitability: From Disks to Active Stars, Planets to Life", and the FWF/NFN subproject, S116 607-N16 "Particle/Radiative Interactions with Upper Atmospheres of Planetary Bodies under Extreme Stellar Conditions". This work was also supported by the Helmholtz Allianz Project "Planetary Evolution and Life" and the International Space Science Institute (ISSI) in Bern, Switzerland.

I also wish to thank H. O. Rucker, W. Baumjohann, H. K. Biernat from the Space Research Institute (IWF) of the Austrian Academy of Sciences (ÖAW) and A. Hanslmeier from the Institute of Geophysics and Meteorology (IGAM) of the University of Graz, Austria for their support of my research activities related to the evolution of planetary atmospheres and its implication for habitability during the past years. Furthermore, I thank Yu. N. Kulikov from the Polar Geophysical Institute (PGI) of the Russian Academy of Sciences in Murmansk, Russian Federation, an innovative contributor to planetary aeronomy, for the foreword and many fruitful scientific discussions during several working visits in the Murmansk and Moscow regions. Special thanks are also given to S. J. Bauer for many fruitful and interesting discussions in the field of comparative planetology during the past decades. He has acted as a valued mentor in many areas related to my studies of the origin and evolution of planetary atmospheres.

Moreover, I acknowledge the support from my colleagues K. G. Kislyakova, M. L. Khodachenko, H. I. M. Lichtenegger, H. Gröller, P. Odert and M. Leitzinger, who worked with me since the past months and years on the various science cases which are covered in this brief monograph. Without their scientific enthusiasm and efforts some of the scientific aspects which are addressed in this work would not have been carried out.

Finally, many thanks to my interdisciplinary cooperators M. Holmström from the Swedish Institute of Space Physics (IRF) in Kiruna, Sweden; N. V. Erkaev from the Institute of Computational Modelling (ICM) of the Siberian Division of the Russian Academy of Sciences in Krasnojarsk, Russian Federation; D. Bisikalo and V. I. Shematovich from the Institute of Astronomy (INASAN) of the Russian

Academy of Sciences in Moscow, Russian Federation; H. Rauer from the Institute of Planetary Research of the German Aerospace Center (DLR) in Berlin, Germany; E. Kallio from the Finish Meteorological Institute (FMI) in Helsinki, Finland; F. Selsis from the Université Bordeaux 1, Bordeaux, France; J.-M.-Grießmeier from the Laboratoire de Physique et Chimie de l'Environnement et de l'Espace (LPC2E) & Observatoire des Sciences de l'Univers en règion Centre (OSUC), Orleans, France; I. Ribas from the Institut d'Estudis Espacials de Catalunya/CSIC, & Campus UAB, Facultat de Ciències, Bellaterra, Spain; and Prof. M. Güdel from the Institute of Astrophysics of the University of Vienna, Austria as well as many colleagues who cooperated with me in the field of planetary atmosphere evolution since the past two decades.

Contents

Chapter 1
Protoatmospheres

For studying the origin and evolution of planetary atmospheres it is important to understand which sources and sinks contributed to their initial formation. As illustrated in Fig. 1.1 the origin of the earliest atmospheres of terrestrial planets can be related mainly on three formation scenarios. The first protoatmosphere scenario is related to the capture and accumulation of a hydrogen- and He-rich nebula gas envelope around the rocky core before the planet ends its accretion. The second formation hypotheses corresponds to dense H_2O/CO_2-rich steam atmospheres, which are catastrophically outgassed during the magma ocean solidification process when the planet's accretion ended. Finally, the third atmosphere producing process is related to the later growth of a secondary outgassed atmosphere by volcanos of H_2O, CO_2, N_2 and other trace gases. It should also be noted that all three hypotheses are strongly linked to the impact history of the individual system and the formation process of the particular planet. Most likely terrestrial planets may obtain their initial atmosphere by a mixture of all three hypotheses but start initially with dense hydrogen-, H_2O-, and CO_2-rich gaseous envelopes.

1.1 Nebula-Based H/He-Rich Gas Envelopes

The massive hydrogen/He atmospheres of gas and ice giants, such as Jupiter, Saturn, Neptune, and Uranus have been formed as a result of unstable gas accretion onto a rocky or icy core [1–5]. Various analytical and numerical studies of these authors indicate that the onset of core instability occurs when the mass of the atmosphere around the core becomes similar to the mass of the core. By studying quasi-static atmospheres of accreting protoplanetary cores for various opacity conditions and planetesimal accretion rates in various locations and orbital distances within a protoplanetary nebula it was found that the type and mass of atmosphere which can be captured around a rocky core depends mainly on the core mass, nebula parameters, and core accretion luminosity [6]. The results of such studies indicate two atmosphere

H. Lammer, *Origin and Evolution of Planetary Atmospheres*,
SpringerBriefs in Astronomy, DOI: 10.1007/978-3-642-32087-3_1,
© The Author(s) 2013

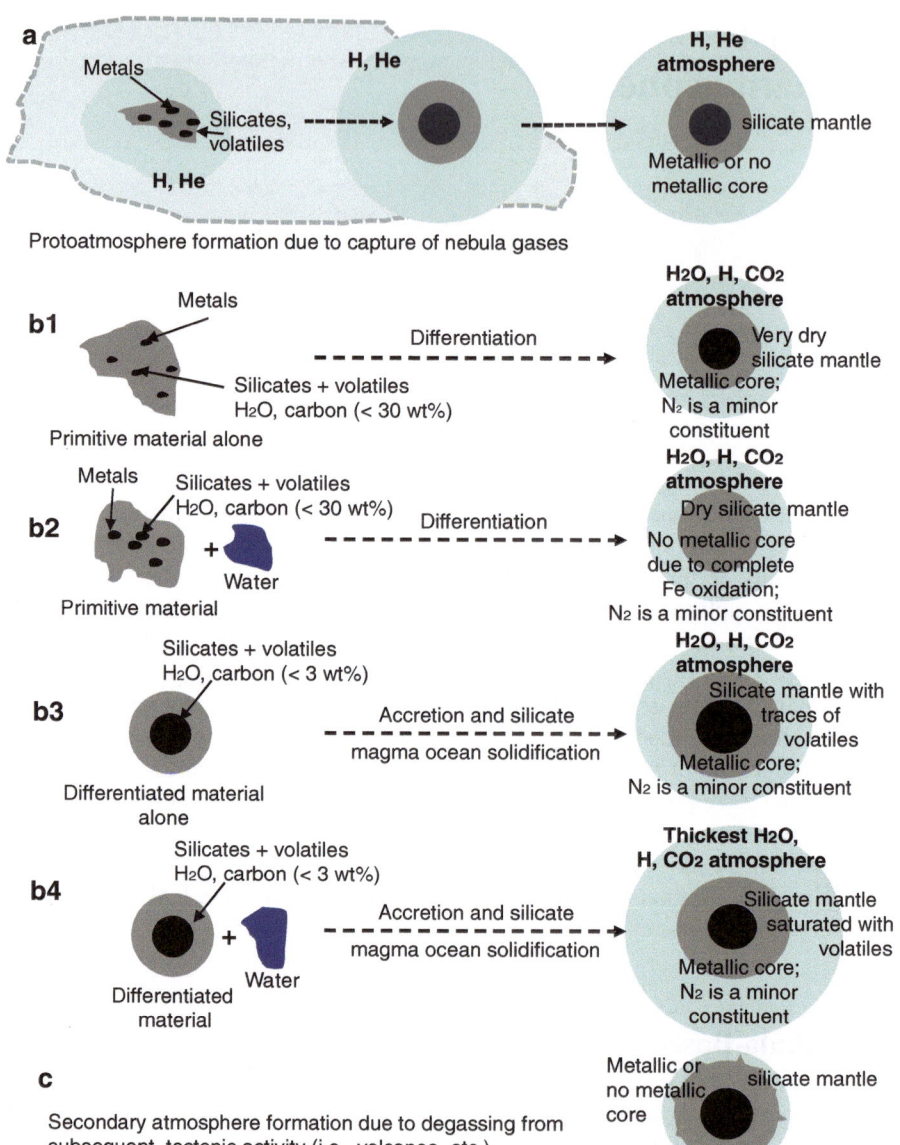

a

Metals

Silicates, volatiles

H, He

H, He

H, He atmosphere

silicate mantle

Metallic or no metallic core

Protoatmosphere formation due to capture of nebula gases

b1

Metals

Silicates + volatiles
H₂O, carbon (< 30 wt%)

Differentiation

Primitive material alone

H₂O, H, CO₂ atmosphere

Very dry silicate mantle

Metallic core; N₂ is a minor constituent

b2

Metals   Silicates + volatiles
H₂O, carbon (< 30 wt%)

+

Water

Differentiation

Primitive material

H₂O, H, CO₂ atmosphere

Dry silicate mantle

No metallic core due to complete Fe oxidation; N₂ is a minor constituent

b3

Silicates + volatiles
H₂O, carbon (< 3 wt%)

Accretion and silicate magma ocean solidification

Differentiated material alone

H₂O, H, CO₂ atmosphere

Silicate mantle with traces of volatiles

Metallic core; N₂ is a minor constituent

b4

Silicates + volatiles
H₂O, carbon (< 3 wt%)

+

Water

Accretion and silicate magma ocean solidification

Differentiated material

Thickest H₂O, H, CO₂ atmosphere

Silicate mantle saturated with volatiles

Metallic core; N₂ is a minor constituent

c

Secondary atmosphere formation due to degassing from subsequent tectonic activity (i.e., volcanos, etc.)

Metallic or no metallic core   silicate mantle

Less dense H₂O, H, CO₂ N₂, atmosphere

◀ **Fig. 1.1** Illustration of various protoatmosphere scenarios for terrestrial planets. There are three primary sources for the formation of terrestrial planetary atmospheres. **a** Illustrates the capture of nebula gases and the accumulation of hydrogen-rich protoatmospheric layers around a rocky planet [2, 3, 6, 7]. **b** Illustrates various possible scenarios of catastrophically outgassed protoatmospheres during accretion and magma ocean solidification [8, 9]. These scenarios result in dense steam atmospheres which can also contain large amounts of CO_2 as well as other volatiles such as CH_4 or NH_3. Nitrogen will be a minor species compared to H_2O and CO_2. **b1** and **b2** illustrates scenarios by accretion and reaction of primordial chondritic materials alone and with additional delivered water, while **b3** and **b4** Illustrates similar scenarios but with accretion and magma ocean processing of differentiated achondritic materials. **c** Illustrates the subsequent formation of secondary atmospheres related to the tectonic activity of a particular terrestrial planet [10] which results also in the outgassing of H_2O, CO_2, N_2, and other minor elements. As soon as the outgassing flux from the planet's interior is higher compared to the escape flux of the volatiles a secondary atmosphere will accumulate

formation processes, one which have an outer convective zone that merges with the surrounding nebula gas and a second one which possesses an isothermal region that decouples the atmosphere from the nebula. Cores of planets within orbit locations ≤ 0.3 AU from solar-like stars have luminosities which favor the first atmosphere type. Cores of massive protoplanets which originate beyond several AU have low luminosities so that massive atmospheres of the second type can be accumulated.

The critical core mass which is needed that the core instability occurs is ~ 5–$20 M_{Earth}$ ($1 M_{Earth} = 5.974 \times 10^{27}$ g) at ~ 0.1–1 AU, and beyond 1 AU about 20–$60 M_{Earth}$ for a nebula opacity of ~ 0.1 cm^2 g^{-1} [6]. The results of such studies suggest that early terrestrial planets including the Earth, which can be considered as lower massive rocky cores should also have accumulated hydrogen-dominated nebula gas envelopes during their formation phase. When the proto-Earth or proto-Venus or other terrestrial planets grow by the accretion of the planetesimals and their mass reached more than $\sim 0.1 M_{Earth}$, a huge amount of nebula gas can be captured toward the protoplanet, so that optically thick, dense hydrogen-rich atmospheres can be formed [4, 11, 12].

Studies indicate that the surface temperature T_s of captured nebula gas can reach values which are $\geq 4,000$ K [11, 12]. Such a surface temperature is high enough that the outer mantle of a protoplanet can melt so that magma oceans originate and a significant amount of noble gases with a solar composition could be incorporated in the planet's interior. These studies indicate that during the early stages of planet formation planetary embryos, which grow into rocky protoplanets in the gaseous disk may accumulate hydrogen-rich gaseous envelopes with an equivalent amount of hydrogen from several tens of Earth oceans (EO_H) up to 1,000 EO_H (1 $EO_H \sim 1.54 \times 10^{23}$ g) [11, 12].

Table 1.1 summarizes the accretion rate $M/\dot{M}$, the mass of a planetary embryo or protoplanet M in units of Earth masses M_{Earth}, T_s in K, accumulated nebula gas in atmospheric mass M_{atm} in g, and hydrogen amounts normalized to one EO_H. One can see from the accumulation rates of nebula gas that growing protoplanets could accumulate the equivalent amount of hydrogen of tens to 1,000 EO_H within a few Myr. However, it is usually assumed that this primordial atmosphere escaped from the Earth after the formation of Jupiter, Saturn and the ice giants and that the solar nebula as a whole was blown away gradually during a few Myr from the Solar System

Table 1.1 Captured hydrogen from the nebula gas during growing proto Earth-type planets [11, 12]

$M/\dot{M}$ (Myr)	M (M_{Earth})	R (R_{Earth})	T_s (K)	M_{atm} (g)	M_{atm} (EO$_H$)
1.0	1.0	~1.0	~4,080	~1.5 × 10^{26}	~1,000
	0.5	~0.85	~2,585	~5.3 × 10^{25}	~350
	0.2	~0.62	~1,310	~2.0 × 10^{25}	~135
	0.1	~0.5	~690	~8.5 × 10^{24}	~55
10.0	1.0	~1.0	~3,980	~3.3 × 10^{26}	~2,200
	0.5	~ 0.85	~2,500	~7.5 × 10^{25}	~500
	0.2	~0.62	~1,240	~2.0 × 10^{25}	~135
	0.1	~0.5	~625	~8.5 × 10^{24}	~55

by a strong solar wind and extreme UV radiation [11]. Critics of the nebula-based protoatmosphere hypothesis expect that the accumulated hydrogen envelopes escape early on, maybe even before the planet ended its accretion and the lifetime of these hydrogen envelopes around young planets during the planetary formation process is most likely limited to a few Myr [5, 13, 14]. Observations show that most gas disks vanish on time scales of ~3 Myr. Disks with ages of ~10 Myr are rather exceptional [5].

If one considers that the accumulated amount of nebula gas given in Table 1.1 decrease after the formation of the gas giants due to the extreme activity of the young star's post T-Tauri phase after ~1–3 Myr the accumulated hydrogen gas mass can be still as heavy as ~2 × 10^{25} g for a nebula gas density ρ_0 ~10^{-19} g cm^{-3} of a protoplanet with a mass close to M_{Earth} [11]. As soon as the nebula gas density reaches that value the particle mean free path becomes larger than the scale hight of the atmosphere and the EUV radiation of the young Sun/Star is able to reach the atmosphere almost freely and heat up the upper regions, so that the gas begins to escape.

Depending on nebula sizes and densities, formation timescales for Jupiter-class gas giants are found close from a few Myr up to ~10 Myr [15] which corresponds more or less to the nebula live time. As discussed before, even after a strong atmospheric loss during a post-T-Tauri phase during that time period Earth-type protoplanets can accumulate hydrogen envelopes with masses of up to ~2×10^{25} g which corresponds to an equivalent amount of ~130 EO$_H$ [11, 12].

These previous results are also supported by a recent study, which was motivated by the discovery of low-density super-Earths in close orbital locations around their host stars. In-situ accretion of hydrogen-rich atmospheres around super-Earths orbiting the Kepler-11 system were studied for investigating if these planets are rock-dominated bodies which are surrounded by thick H/He gas envelopes [7]. From the results of these recent studies one can conclude that hydrogen-rich atmospheres can indeed accumulate around these low-density super-Earths if the disk dissipated slowly or they originated in cool environments [7]. From these studies which are in agreement with the observation of the Kepler-11 low-density super-Earths one obtains for their mass range between ~2–10 M_{Earth} accumulated hydrogen atmosphere equivalents of $\sim$770–7,700 EO$_H$ if one assumes a $\sim$1 % ratio of $M_{ath}/(M_{ath}+M_{rock})$ and $\sim$3,880–38,800 EO$_H$ if one assumes a $\sim$10 % ratio of $M_{ath}/(M_{ath}+M_{rock})$, respectively.

Depending on the assumed EUV heating efficiency for hydrogen-rich atmospheres which lies within an uncertainty range of $\sim$15–60 % and observation-based soft X-ray and corresponding EUV flux values from solar proxies the escape of hydrogen atoms since a $\sim$3 Myr young proto-Earth can be estimated of $\sim$8–25 EO$_H$ during the most active solar/stellar phase of $\sim$500 Myr after the host star's origin [16]. From these estimates one can expect that young terrestrial planets, including the early Earth could have been surrounded with hydrogen envelopes to equivalent amounts of at least a few EO$_H$. If this was indeed the case then one can see from Table 1.1, that the environmental conditions of nebula-based hydrogen-rich protoatmospheres may not have reached surface temperatures, which are required for the formation of magma oceans so that solar noble gases could not be incorporated directly into the mantle material.

However, the capture of nebula gases around low mass rocky planets is a complex process which is strongly linked to the disk dissipation time, the formation process of and orbital locations of the system's gas giants or more massive planets so that its relevance for low mass terrestrial planets is most likely different for various systems and therefore not well understood.

To overcome the problem of a nebula-based protoatmosphere for the explanation of Earth's solar-noble gas content one could suggest that these noble gases where incorporated into the Earth from accretion of earlier objects [17] such as large planetesimals or planetary embryos. Another interesting possibility that solar noble gases might be explained by drop of them into early formed planetesimals [18]. With the evidence that the young Sun was much more active compared to the modern Sun [19–21] one can expect that inner Solar System planetesimals should have been implanted by solar wind related noble gases.

1.2 Catastrophically Outgassed Steam Atmospheres

Terrestrial planets, with silicate mantles and metallic cores, will obtain the majority of their initial water and carbon compounds during the formation process. Decades ago terrestrial planets were thought to have accreted from primordial undifferentiated disk material similar as chondritic meteorites [22]. In such a scenario, terrestrial planets were formed in a volatile-free and a volatile-rich component so that the reduced metallic core and the oxidized mantle and surface could be explained. A two-component accretion process was also assumed [23], where one component originated closer to the Sun and the second oxidized component formed at larger orbital locations. Present day terrestrial planet formation model simulations indicate that terrestrial planets originate from differentiated large planetary embryos with sizes of several hundred to a few thousand kilometers [14, 24, 25]. After the evaporation of the nebula gas due to the extreme EUV phase of the young Sun/Star the protoplanets continue to grow through the capture and collisions of remaining large planetesimals and planetary embryos [14, 24–27].

In modern terrestrial planet simulations, planets with Mars-type masses up to super-Earths are usually formed within ≤ 100 Myr. The initial water inventory of newly created planets depend mainly on the configuration of the protoplanetary disk, the location of the ice line, the orbit location, mass and eccentricity of additional planets, and the impact history of asteroids and comets. In the case of the Earth the buoyant hydrous phases were produced by the magma ocean and icy impactors which originated from beyond the ice line and kept it permanently molten or molten in a series of molten phases for a few tens of Myr [8, 12, 14, 15, 25, 28–32].

The range of possible bulk compositions for terrestrial planets in the Solar System reaches from an extreme end-member, which contains a mixed rocky bulk composition with iron metal meteorites but no water, or mixtures of metallic iron, silicate rock, and unconstrained volumes of icy planetesimals and planetary embryos from beyond the ice line [9]. Based on studies of meteorites, asteroids, and comets a wide range of initial water inventories between ~ 0.01–3 weight percent (wt%) in the bulk silicate of terrestrial planets can be expected [8, 33–35]. The pressure gradient in the magma ocean of the accreted planet controls the solidification process and is related to the planetary mass. Because most minerals that solidify from a bulk planetary magma composition are denser compared to the coexisting magmatic material, the solidification process proceeds from the bottom upward to the surface with increasing amounts of magma lying above solid silicate minerals [8, 9, 32, 35, 36]. The mantle solidification of a magma ocean is a fast process and ends at $\sim 10^5$ years for Earth-size planets with low volatile contents and at ≤ 3 Myr for planets with higher volatile contents and magma ocean depths of $\leq 2{,}000$ km [8, 35]. During the magma ocean solidification process, H_2O and CO_2 molecules can enter the solidifying minerals in relative low quantities [8, 9]. As a result the H_2O/CO_2 volatiles will degas into dense steam atmospheres [8, 9, 34, 35].

Figure 1.2 shows the estimated surface pressures of catastrophically outgassed steam atmospheres as a function of planetary mass and initial water content [35]. The shaded area corresponds to the range of water content of differentiated meteorites which can be considered as the building blocks of the terrestrial planets. The initial bulk water content in the whole mantle of the magma ocean is shown in the dashed lines, while the solid lines indicate the water content of the final 1 vol% of magma ocean liquids when the solidification process is mainly finished. Depending on the initial conditions one can see that steam atmospheres which are produced by degassing from magma oceans can reach initial surface pressure values from hundreds to tens of thousands of bar [8, 9, 35]. In case of the early Earth, if all the expected surface and upper mantle H_2O/CO_2 content could be converted into a primitive protoatmosphere, a water dominated steam atmosphere with ~ 560 bar of H_2O and about ~ 100 bar of CO_2 could build up [8].

Table 1.2 shows the range of expected outgassing products for the early Earth. The corresponding surface pressure value depends on the estimated range of the initial volatile content, the depth of Earth's magma ocean and its solidification time. This inventory is also called after the American geologist Rubey, "Rubey inventory" [10].

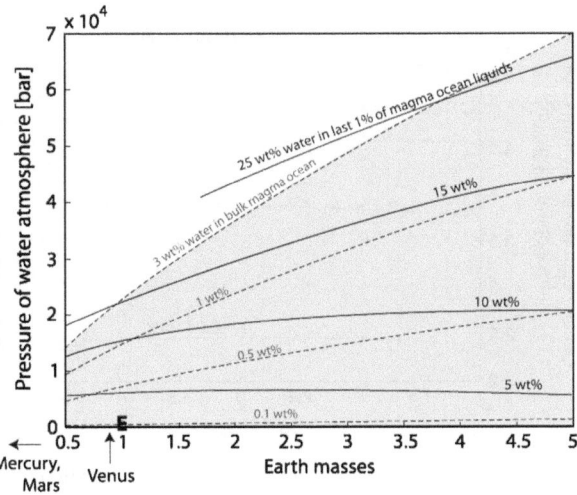

Fig. 1.2 Surface pressure of steam atmospheres, which are produced by catastrophically degassing during the solidification of magma oceans as a function of planetary mass and initial water content. The *shaded area* corresponds to the possible water content of achondritic meteorites. The *dashed lines* correspond to an initial bulk water content of final 1 vol% of magma ocean liquids before the solidification ends (after [35])

Table 1.2 Terrestrial outgassing products and partial surface pressure in units of bar, expected for the early Earth [8, 10, 37]

Species	H_2O	CO_2	Cl	N_2	S	Ar, etc.
P (bar)	~250–550	~50–70	~7	≥1	≥0.5	0.03

Table 1.3 Surface pressure of outgassed water dominated steam atmospheres, in units of bar and in Earth Ocean's (EOs) for a planet with $1 M_{Earth}$ and various initial bulk water inventories inside a 2,000 km deep magma ocean with water contents in weight percent (wt%) [8, 9, 35]

	0.05 wt%	0.1 wt%	0.5 wt%	1 wt%	3 wt%
$1 M_{Earth}$	~250 bar	~500 bar	~7,000 bar	~15,000 bar	~23,000 bar
	~1 EO	~2 EOs	~28 EOs	~60 EOs	~92 EOs

Various degassing and protoatmosphere formation scenarios during accretion for a wide range of atmospheric masses and composition of exoplanets within a range between ~1–30 M_{Earth} have been modeled [9, 35]. By using primitive and differentiated meteorite compositions these authors found that degassing alone can produce a wide range of planetary atmospheres which range from ≤1 % of a planet's mass up to ~6 mass% of hydrogen, ~20 mass% of H_2O, and ~5 mass% of carbon compounds. Hydrogen-rich atmospheres can also be outgassed due to oxidation of metallic Fe with H_2O.

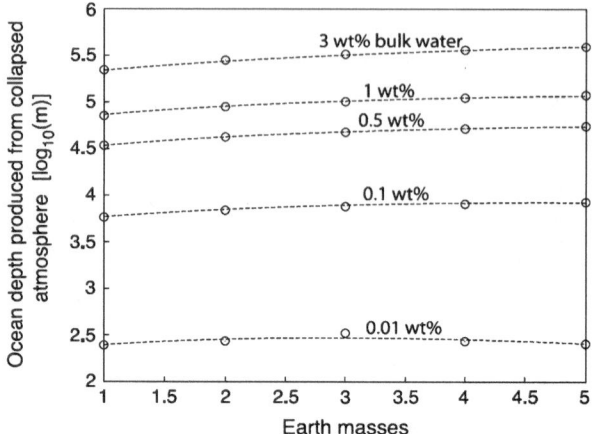

Fig. 1.3 Water ocean depth from collapsed steam atmospheres as a function of planetary mass and initial wt% of bulk water in the magma ocean (after [35])

As shown in Fig. 1.2 and Table 1.3, for a higher initial volatile content, these models can produce early atmospheres with surface pressures up to several 10^4 bar. From Fig. 1.2 one can also see that a super-Earth with $3M_{Earth}$ that starts with 3 wt% H_2O in the whole mantle of its magma ocean will end up with about $\sim$1 wt% of magma ocean liquids which contain between $\sim$15–25 % water. If the initial H_2O inventory is >1–3 wt%, which can be expected for super-Earths, supercritical water, which is an environment where H_2O exists in liquid and gas form together can directly originate [9, 35]. In this supercritical phase liquid water expands and molecules become steam and rise above the liquid. The vapor density consequently increases, while the density of the liquid decreases. For an Earth-like planet, surface temperatures near the supercritical point of H_2O can be reached within several millions to tens of Myr after the solidification of a magma ocean. These timescales agree also with studies by several researchers who investigated the early stages of accretion and impacts and expect that due to thermal blanketing temperatures $\geq$1,500 K could keep the volatiles of the protoatmosphere in vapor phase within several tens of Myr [3, 11, 12, 32, 38–40].

As it is shown in Fig. 1.3 if the planet orbits within the habitable zone of its host star at the end of such cooling epochs the water vapor which did not escape to space can condense and form deep oceans. For super-Earths with dense steam atmospheres up to several 10^4 bar this timescale may last even a few hundreds of Myr [8, 35]. Depending on the heating of the magma ocean and the impact history of the particular planet, in the case of Earth, it is expected that surface temperatures near the supercritical point of H_2O are reached within tens of million years. As illustrated in the P–T phase diagram of water in Fig. 1.4, one can see that if the surface temperature of a planet cools below the critical point of $\sim$647 K related to a

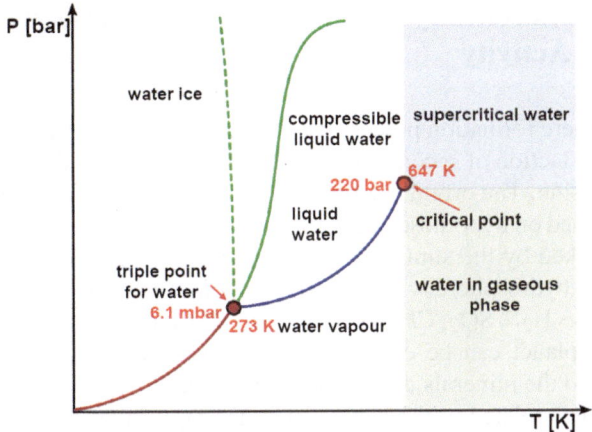

Fig. 1.4 Pressure-temperature phase diagram of water. If the temperature and pressure reaches values which are larger than the critical point ($T \sim 647$ K, $P \sim 220$ bar) water exists in supercritical state. The combination of pressure and temperature at which liquid water, ice, and vapor can coexist in an equilibrium occurs at the triple point which corresponds to 273.16 K and a partial vapor pressure of ~ 6.1 mbar

pressure of ~ 220 bar the supercritical environment can collapse into a liquid water ocean [35].

For early Earth, it is suggested [8] in agreement with EUV-powered protoatmosphere escape studies [16], that initial magma ocean water inventories $\gg 0.1$ wt% were unlikely, because they would have implied initial atmosphere pressures of hundreds to several thousands of bars and times related to cooling of clement surface conditions may have lasted hundreds of Myr. Such long cooling times in combination with high surface pressures do not agree with the evidence for liquid water on Earth's surface discovered in zircons ~ 4.4 Gyr ago [41].

From the brief discussions of the protoatmosphere formation scenarios, one understands that the initial conditions, which are related to the formation of the earliest atmospheres of terrestrial planets determine if an Earth-like planet may evolve to a class I, II [42], or a class V habitat. If the early Earth would have obtained slightly more material from water-rich planetesimals, its CO_2 content would have been much higher and Earth's oceans could have been tens to hundreds of kilometers deep (see also Fig. 1.3). Such environmental conditions would have resulted in a globally covered water world [43, 44] which is surrounded by a Venus-type dense CO_2 atmosphere and a hydrogen envelope. The corresponding greenhouse effect would have produced a hot evaporating upper ocean layer. From this complex connection between the formation of terrestrial planets, the impact history of the planetary system and the related protoatmosphere formation one can expect that the Universe will be populated with terrestrial planets, which may have a wide variety of initial atmospheres with different water and CO_2 contents.

1.3 Degassed Secondary Atmospheres from Subsequent Tectonic Activity

A third atmosphere formation process which acts on every terrestrial planet to some extent is the production of secondary atmospheres due to tectonic processes such as volcanic outgassing. The origin of secondary atmospheres from subsequent tectonic activities are based on hypotheses which were suggested decades ago [10, 45]. Their theories are backed by the similarity in the chemical composition between Earth's atmosphere and hydrosphere with the composition of volcanic gases such as H_2O, CO_2, and in traces H_2S, SO_2, CH_4, NH_3, HF, H_2, CO, Ar, etc. The "Rubey inventory" of a terrestrial planet can be considered as the volatile fraction which could be incorporated into the minerals during the magma ocean solidification process [8].

Several models related to the history of degassing of the early Earth, which are based on the hypothesis that a relatively undegassed lower mantle supplied volatiles up to the upper mantle which were then released into the surrounding environment were reviewed in detail more then a decade ago [5]. This hypothesis is supported by the isotopic composition and concentration of ^{40}Ar which provides an evidence that gases have been added to the atmosphere from the planet's interior over geological timescales [32, 46, 47]. Xe isotope studies related to the mantle and atmosphere of the Earth indicate that the planetary rock and the atmosphere separated from each other during an early stage [5, 47]. However, it is highly probable that part of the Xe in Earth's early atmosphere was added via impacts after the degassing process [31, 48–50] so that differences between elemental and atomic abundances of mantle noble gases relative to the atmosphere can also be explained by heterogenous atmospheric accretion [50–52].

Additionally to volcanic outgassing, atmospheric volatiles were also delivered to terrestrial planets by impacts. It is certainly beyond the scope of this brief monograph to address the complex subject of secondary atmosphere formation in detail and its connection to the previously discussed nebula and/or magma ocean-related degassed protoatmospheres for Venus, Earth or terrestrial planets in general. On the other hand, there are strong indications that a secondary outgassed CO_2 atmosphere on early Mars may have played an important role [53, 54]. A brief discussion on it's importance for the evolution of the early Mars atmosphere is discussed below.

The question if and when the martian atmosphere was dense enough during it's history to allow liquid water to be stable on the surface is closely related to the planet's habitability [42, 54–59]. When the accretion and core formation on early Mars ended, the young planet was also most likely also covered by a global magma ocean. During the solidification of this magma ocean a dense steam atmosphere could built up [8, 54, 60]. Estimates of the surface pressure of such a steam-type protoatmosphere yield ≤ 100 bar [8, 60], and the corresponding atmosphere was most likely dense enough to significantly reduce magma ocean cooling [8]. There is observational evidence from solar proxies with younger age compared to the present Sun, that during the early history of the Solar System the EUV flux was up to ~ 100 times larger as it is today [16, 19, 20, 42, 54, 59, 61, 62]. As a consequence, the temperatures

Table 1.4 Modeled total CO_2 partial surface pressure and equivalent global water layer (EGL_{H_2O}) originating from tectonic outgassing as a function of Mars' age.

t_{Mars} (Gyr)	0.1	0.25	0.5	0.75	1.0	1.5
GM_{CO_2} (bar)	~0.25	~0.6	~0.8	~1	~1	~1
PM_{CO_2} (bar)	≤0.1	~0.25	~0.35	~0.5	~0.65	~0.8
GM EGL_{H_2O} (m)	~13.5	~34	~50	~57	~61	~61
PM EGL_{H_2O} (m)	~2.5	~5.5	~6.8	~10	~13.5	~15

The upper values correspond to a model which considers mantle melting in a global melt channel (GM), whereas the lower values correspond to a model considering melting in mantle plumes (PM) which covers only a small fraction of the planetary surface [54]

Table 1.5 Secondary outgassed CO_2 partial surface pressure in units of bar as function of iron-wüstite (IW) buffer and surface fraction of the melt channel (f_p) by assuming that ~400 Myr after the planets origin the outgassing flux is assumed larger than the escape flux [54]

IW and f_p scenarios	CO_2 (mbar)
IW=1; f_p=0.01	~600
IW=1; f_p=1	~250
IW=0; f_p=0.01	~110
IW=0; f_p=1	~50

in the upper atmosphere of early Mars significantly increased [53, 54, 62, 63]. This strong thermospheric EUV-heating results in an expansion of the upper atmosphere, leading to efficient atmospheric escape through thermal and non-thermal loss of heavy, neutral atoms, such as O, C, and N atoms so that an early CO_2-rich upper atmosphere could not have been maintained [53, 54]. After the protoatmosphere escaped, mantle outgassing most likely have built up a secondary CO_2 atmosphere which is shown in Table 1.4 of about ≤1 bar or less ~3.8–4.3 Gyr ago [54, 64]. If one considers that the escape flux of CO_2 was less than the outgassed flux from the planet's interior ~400 Myr after the planet's origin, a secondary CO_2 atmosphere of ~600 mbar could build up if one assumes a global melt channel [54]. As shown in Table 1.5 by considering different iron-wüstite (IW) IW-buffers or melt channel values one obtains a lower surface pressure. It is important to note that additionally to the secondary outgassed atmosphere significant amounts of H_2O and carbon may have been delivered also later by comets.

According to various terrestrial planet formation models [30, 31] an equivalent amount of water of ~0.1 EOs, that corresponds to an ~300 m deep EGL of water, could have provided to Earth by comets during the few 100 Myr following main accretion. For early Mars, a fraction of such impact delivered water, if not all of it was lost due to the high thermal escape rate [53, 54], could have been remained and is most likely stored in the planet's crust. Thus, after the formation of a secondary produced CO_2 atmosphere ~4 Gyr ago the atmospheric surface pressure may have been also modified by impact erosion and delivery, CO_2 adsorption in the regolith, sequestration in carbonate rocks and storage in ice [54, 65].

However, it is important to note that the N_2/CO_2 ratio in the martian atmosphere of 2.6 %/95 %~0.0027 is only slightly higher than Venus' ratio of 1.8 %/98 %~0.018 but orders of magnitude lower compared to Earth's ratio of 78 %/ 0.03 % ~2,600, where most of the CO_2 is stored in the lithosphere in the form of carbonates. Thus, the present atmospheres of Mars and Venus are in a good agreement with the outgassed N_2/CO_2 ratio of the "Rubey inventory" of ~0.015 before the sequestering of CO_2 in the form of carbonates on Earth [37, 66]. This fact indicates that most likely only a small fraction of martian CO_2 may have transformed into carbonates.

References

1. Stevenson, D.J.: Formation of the giant planets. Planet. Space Sci. **30**, 755–764 (1982)
2. Wuchterl, G.: The critical mass for protoplanets revised—massive envelopes through convection. Icarus **106**, 323–334 (1993)
3. Mizuno, H.: Formation of the giant planets. Prog. Theor. Phys. **64**, 544–557 (1980)
4. Ikoma, M., Nakazawa, K., Emori, H.: Formation of giant planets: dependences on core accretion rate and grain opacity. ApJ **537**, 1013–1025 (2000)
5. Halliday, A.N.: The origin of the earliest history of the Earth. Treatise Geochem. **1**, 509–557 (2003)
6. Rafikov, R.R.: Atmospheres of protoplanetary cores: critical mass for nucleated instability. ApJ **648**, 666–682 (2006)
7. Ikoma, M., Hori, Y.: In-situ accretion of hydrogen-rich atmospheres on short-period super-Earths: implications for the Kepler-11 planets. ApJ, arXiv:1204.5302v1 (2012) (submitted)
8. Elkins-Tanton, L.T.: Linked magma ocean solidification and atmospheric growth for Earth and Mars. Earth Planet. Sci. Lett. **271**, 181–191 (2008)
9. Elkins-Tanton, L., Seager, S.: Ranges of atmospheric mass and composition of super-Earth exoplanets. ApJ **685**, 1237–1246 (2008)
10. Rubey, W.W.: Geological history of seawater. Bull. Geol. Soc. Am. **62**, 1111–1148 (1951)
11. Hayashi, C., Nakazawa, K., Mizuno, H.: Earth's melting due to the blanketing effect of the primordial dense atmosphere. Earth Planet. Sci. Lett. **43**, 22–28 (1979)
12. Matsui, T., Abe, Y.: Impact-induced atmospheres and oceans on Earth and Venus. Nature **322**, 526–528 (1986)
13. Najita, J.R., Strom, S.E., Muzerolle, J.: Demographics of transition objects. Mon. Not. R. Astron. Soc. **378**, 369–378 (2007)
14. Lunine, J.I., O'Brien, D.P., Raymond, S.N., Morbidelli, A., Qinn, T., Graps, A.L.: Dynamical models of terrestrial planet formation. Adv. Sci. Lett. **4**, 325–338 (2011)
15. Alibert, Y., Broeg, C., Benz, W., Wuchterl, G., Grasset, O., Sotin, C., Eiroa, C., Henning, T., Herbst, T., Kaltenegger, L., Léger, A., Liseau, R., Lammer, H., Beichman, C., Danchi, W., Fridlund, M., Lunine, J., Paresce, F., Penny, A., Quirrenbach, A., Röttgering, H., Selsis, F., Schneider, J., Stam, D., Tinetti, G., White, G.J.: Origin and formation of planetary systems. Astrobiology **10**, 19–32 (2007)
16. Lammer, H., Kislyakova, K.G., Odert, P., Leitzinger, M., Schwarz, R., Pilat-Lohinger, E., Kulikov, Yu.N., Khodachenko, M.L., Güdel, M., Hanslmeier, A.: Pathways to Earth-like atmospheres: extreme ultraviolet (EUV)-powered escape of hydrogen-rich protoatmospheres. Orig. Life Evol. Biosph. **41**, 503–522 (2012)
17. Farley, K.A., Poreda, R.: Mantle neon and atmospheric contamination. Earth Planet Sci. Lett. **114**, 325–339 (1992)
18. Podosek, F.A.: Noble gases. Tr. Geochem. **1**, 381–405 (2003)
19. Güdel, M., Guinan, E.F., Skinner, S.L.: The X-ray sun in time: a study of the long-term evolution of coronae of solar-type stars. ApJ **483**, 947–960 (1997)

20. Ribas, I., Guinan, E.F., Güdel, M., Audard, M.: Evolution of the solar activity over time and effects on planetary atmospheres: I. High-energy irradiances (1–1700Å). ApJ **622**, 680–694 (2005)
21. Güdel, M.: The Sun in time: activity and environment. Living Rev. Sol. Phys. **4**(3) (2007)
22. Ringwood, A.E.: Origin of the Earth and Moon, p. 307. Springer, New York (1979)
23. Wänke, H., Dreibus, G.: Chemical composition and accretion history of terrestrial planets. Philos. Trans. R. Soc. A **325**, 545–557 (1988)
24. Kokubo, E., Ida, S.: Formation of protoplanets from planetesimals in the solar nebula. Icarus **143**, 15–27 (2000)
25. Raymond, S.N., Quinn, T., Lunine, J.I.: Making other Earths: dynamical simulations of terrestrial planet formation and water delivery. Icarus **168**, 1–17 (2004)
26. Weidenschilling, S.J.: The distribution of mass in the planetary system and solar nebula. Astrophys. Space Sci. **51**, 153–158 (1977)
27. Wetherill, G.W.: Accumulation of terrestrial planets and implications concerning lunar origin. In: Hartmann, W.K., Phillips, R.J., Taylor, G.J. (eds.) Origin of the Moon Lunar and Planet. pp. 519–550, University Arizona press, Chicago (1986)
28. Anders, E., Owen, T.: Mars and Earth—origin and abundance of volatiles. Science **198**, 453–465 (1977)
29. Zahnle, K.J., Kasting, J.F., Pollack, J.B.: Evolution of a steam atmosphere during Earth's accretion. Icarus **74**, 62–97 (1988)
30. Lunine, J.I., Chambers, J., Morbidelli, A., Leshin, L.A.: The origin of water on Mars. Icarus **165**, 1–8 (2003)
31. Morbidelli, A., Chambers, J., Lunine, J.I., Petit, J.M., Robert, F., Valsecchi, G.B., Cyr, K.: Source regions and timescales for the delivery of water to Earth. Meteorit. Planet. Sci. **35**, 1309–1320 (2000)
32. Albarède, F., Blichert-Toft, J.: The split fate of the early Earth, Mars, Venus and Moon. CR Geosci. **339**, 917–927 (2007)
33. Jarosewich, E.: Chemical analysis of meteorites: a combination of stony and iron meteorite analyses. Meteoritics **25**, 323–337 (1990)
34. Liu, L.-G.: The inception of the oceans and CO_2-atmosphere in the early history of the Earth. Earth Planet Sci. Lett. **227**, 179–184 (2004)
35. Elkins-Tanton, L.T.: Formation of water ocean on rocky planets. Astrophys. Space Sci. **332**, 359–364 (2011)
36. Solomatov, V.S.: Fluid dynamics of a terrestrial magma ocean. In: Origin of the Earth and the Moon. University Arizona Press, Tucson, pp 323–338 (2000)
37. Bauer, S.J.: Über die Entstehung der Planetenatmosphären. Arch. Met. Geoph. Biokl. Ser. A textbf27, 217–232 (1978)
38. Abe, Y.: Thermal and chemical evolution of the terrestrial magma ocean. Phys. Earth Planet. Inter. **100**, 27–39 (1997)
39. Zahnle, K.J., Kasting, J.F.: Mass fractionation during transonic escape and implications for loss of water from Mars and Venus. Icarus **68**, 462–480 (1986)
40. Hunten, D.M., Pepin, R.O., Walker, J.C.G.: Mass fractionation in hydrodynamic escape. Icarus **69**, 532–549 (1987)
41. Valley, A.M., Peck, W.H., King, E.M., Wilde, S.A.: A cool early Earth. Geology **30**, 351–354 (2002)
42. Lammer, H., Bredehöft, J.H., Coustenis, A., Khodachenko, M.L., Kaltenegger, L., Grasset, O., Prieur, D., Raulin, F., Ehrenfreund, P., Yamauchi, M., Wahlund, J.-E., Grießmeier, J.-M., Stangl, G., Cockell, C.S., Kulikov, Yu.N, Grenfell, L., Rauer, H.: What makes a planet habitable? Astron. Astrophys. Rev. **17**, 181–249 (2009)
43. Léger, A., Selsis, F., Sotin, C., Guillot, T., Despois, D., Mawet, D., Ollivier, M., Labèque, A., Valette, C., Brachet, F., Chazelas, B., Lammer, H.: A new family of planets? "Ocean-planets". Icarus **169**, 499–504 (2004)

44. Selsis, F., Chazelas, B., Bordé, P., Ollivier, M., Brachet, F., Decaudin, M., Bouchy, F., Ehrenre-ich, D., Grießmeier, J.-M., Lammer, H., Sotin, C., Grasset, O., Moutou, C., Barge, P., Deleuil, M., Mawet, D., Despois, D., Kasting, J.F., Léger, A.: Could we identify hot ocean-planets with CoRoT, Kepler and Doppler velocimetry? Icarus **191**, 453–468 (2007)

45. Brown, H.: Rare gases and the formation of the Earth's atmosphere, In: Kuiper, G.P (ed.) The Atmospheres of the Earth and Planets, pp 258–266. University of Chicago Press, Chicago (1949)

46. Porcelli, D., Wasserburg, G.J.: Mass transfer of Helium, Neon, Argon, and Xenon through a steady-state upper mantle. Geochim. Cosmochim. Acta. **59**, 4921–4937 (1995)

47. Allègre, C.J., Hofmann, A.W., O'Nions, R. K.: The argon constraints on mantle structure. Geophys. Res. Lett. **23**, 3555–3557 (1996)

48. Javoy, M.: The birth of the Earth's atmosphere: the behaviour and fate of its major elements. Chem. Geol. **147**, 11–25 (1998)

49. Owen, T-, Bar-Nun, A.: Volatile contributions form icy planetesimals, In: Camp, R.M., Righter, K (eds.) Origin of the Earth and Moon, pp 459–471. University of Arizona Press, Tucson (2000)

50. Dauphas, N.: The dual origin of the terrestrial atmosphere. Icarus **165**, 326–339 (2003)

51. Marty, B.: Neon and Xenon isotopes in MORB: implications for Earth-atmosphere evolution. Earth Planet. Sci. Lett. **94**, 45–56 (1989)

52. Caffee, M.W., Hudson, G.B., Velsko, C., Huss, G.R., Alexander Jr, E.C., Chivas, A.R.: Pri-mordial noble gases from Earth's mantle: identification of a primitive volatile component. Sci. **285**, 2115–2118 (1999)

53. Tian, F., Kasting, J.F., Solomon, S.C.: Thermal escape of carbon from the early martian at-mosphere. Geophys. Res. Lett. **36**(2), CiteID L02205 (2009)

54. Lammer, H., Chassefière, E., Karatekin, Ö, Morschhauser, A., Niles, P.B., Mousis, O., Grott, M., Gröller, H., Hauber, E., Pham, L.B.S.: Outgassing history and escape of the martian atmosphere and water inventory. Space Sci. Rev. submitted (2012)

55. Kasting, J.F.: CO_2 condensation and the climate of early Mars. Icarus **94**, 1–13 (1991)

56. Kasting, J.F.: The early Mars climate question heats up. Science **278**, 1245 (1997)

57. Forget, F., Pierrehumbert, R.T.: Warming early Mars with carbon dioxide clouds that scatter infrared radiation. Science **278**, 1273–1276 (1997)

58. Lammer, H., Stumptner, W., Bauer, S.J.: Loss of H and O from Mars: implications for the planetary water inventory. Geophys. Res. Lett. **23**, 3353–3356 (1996)

59. Lammer, H., Kasting, J.F., Chassefière, E., Johnson, R.E.: Kulikov, Yu.N., Tian, F.: Atmospheric escape and evolution of terrestrial planets and satellites. Space Sci. Rev. **139**, 399–436 (2008)

60. Hirschmann, M.M., Withers, A.C.: Ventilation of CO_2 from a reduced mantle and consequences for the early Martian greenhouse. Earth Planet. Sci. Lett. **270**, 147–155 (2008)

61. Zahnle, K.J., Walker, J.C.G.: The evolution of solar ultraviolet luminosity. Rev. Geophys. **20**, 280–292 (1982)

62. Lammer, H.: Kulikov, Yu.N, Lichtenegger, H.I.M.: Thermospheric X-ray and EUV heating by the young Sun on early Venus and Mars. Space Sci. Rev. **122**, 189–196 (2006)

63. Kulikov, Yu.N: Lammer, H., Lichtenegger, H.I.M., Penz, T., Breuer, D., Spohn, T., Lundin, R., Biernat, H.K.: A comparative study of the influence of the active young Sun on the early atmospheres of Earth, Venus and Mars. Space Sci. Rev. **129**, 207–243 (2007)

64. Grott, M., Morschhauser, A., Breuer, D., Hauber, E.: Volcanic outgassing of CO_2 and H_2O on Mars. Earth Planet. Sci. Lett. **308**, 391–400 (2011)

65. Chassefière, E., Leblanc, F.: Constraining methane release due to serpentinzation by the ob-served D/H ratio on Mars. Earth Planet. Sci. Lett. **310**, 262–271 (2011)

66. Bauer, S.J.: Origin of planetary atmospheres and their role in the evolution of life. In: Lacoste, H. (ed.) Proceedings of the Second European Workshop on Exo-Astrobiology , ESA SP-518, pp. 21–24 (2002)

Chapter 2
Evolution of the Solar/Stellar Radiation and Plasma Environment

The evolution of planetary atmospheres can only be understood if one considers that the radiation and particle environment of the Sun or a planet's host star changed during their life time. The magnetic activity of solar-type stars declines steadily during their evolution on the Zero-Age-Main-Sequence (ZAMS). According to the solar standard model, the Sun's photospheric luminosity was ∼30 % lower ∼4.5 Gyr ago when the Sun arrived on the ZAMS compared to present levels. The observed faster rotation of young stars is responsible for an enhanced magnetic activity and related heating processes in the chromosphere, X-ray emissions are ≥1,000, and EUV, and UV ∼100 and ∼10 times higher compared to today's solar values. Moreover, the production rate of high-energy particles is orders of magnitude higher at young stars, and from observable stellar mass loss-activity relations one can also expect a much stronger solar/stellar wind during the active stellar phase. The interaction with these high-energy radiation and the solar/stellar wind plasma interaction with upper planetary atmospheres modifies the thermospheric density and temperature structure and effects finally the whole planetary evolution due to thermal and non-thermal atmospheric escape processes [1].

2.1 Solar/Stellar-Atmospheric Temperature Relations

The effective temperature T_{eff} of a planetary surface in the absence of an atmosphere can be derived from the energy balance between the optical/near infrared emission which irradiates the planet and the mid-infrared (IR) thermal radiation that is lost to space

$$T_{eff} = \left[\frac{S(1 - A)}{4\sigma_B \varepsilon} \right]^{\frac{1}{4}}, \qquad (2.1)$$

with S the flux density of incoming solar/stellar electromagnetic radiation per unit area that would be incident on a plane perpendicular to the rays, at the planet's

H. Lammer, *Origin and Evolution of Planetary Atmospheres*,
SpringerBriefs in Astronomy, DOI: 10.1007/978-3-642-32087-3_2,
© The Author(s) 2013

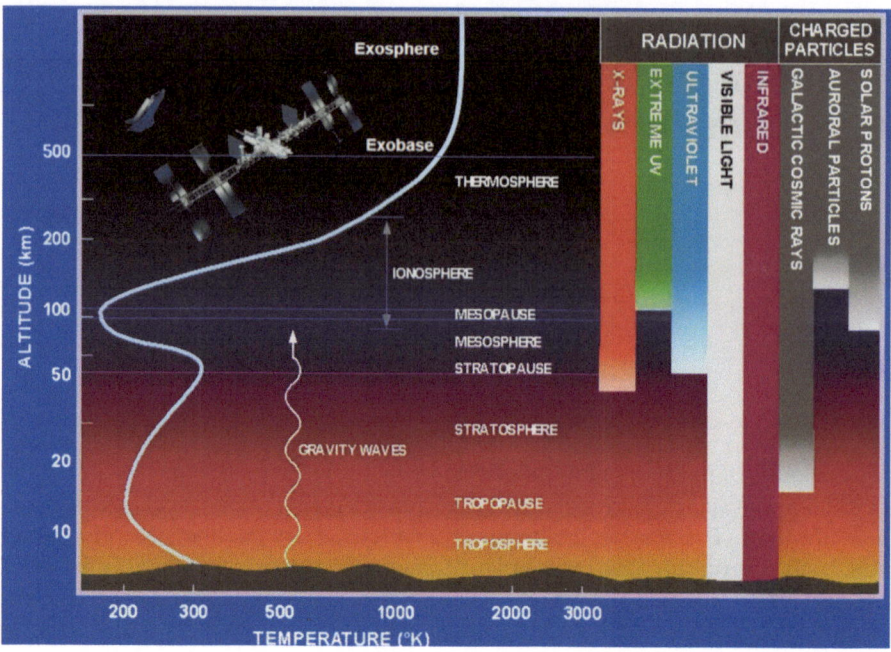

Fig. 2.1 Illustration of the Earth's temperature profiles, atmospheric regions, and incoming solar radiation. Longer wavelengths reach the surface and control a planet's climate, while shorter wavelengths are absorbed in the upper atmosphere and control the thermosphere and exobase temperature

distance in AU (for the Earth: solar constant), σ_B the Stefan-Boltzmann constant, the planet's albedo A and the surface mid-IR emissivity ε. If a planet has an atmosphere, T_{eff} can be higher due to the presence of greenhouse gases. Greenhouse gases such as H_2O, CH_4, CO_2 are nearly transparent to the incident optical and near-infrared (IR) radiation but strongly absorb re-emitted thermal radiation in the mid-IR. The Earth's present day atmosphere leads to a greenhouse effect which results in a daily average surface temperature of about $\sim$288 K [2].

The optical and infrared emissions of young stars largely control the lower regions of a planetary atmosphere (e.g., troposphere, mesosphere) and their related climate. The more energetic shorter wavelength emissions of X-ray, soft X-ray (SXR), EUV photons as well as the interaction with high-energy particles, the stellar wind, or coronal mass ejections (CMEs) lead to heating and expansion of the upper atmospheres [3, 4].

As illustrated in Fig. 2.1 the heat production by the incoming solar/stellar EUV radiation in a planet's upper atmosphere is balanced by the divergence of the conductive heat flux of the EUV radiation. In the so-called thermosphere the short wavelength radiation of is absorbed and transferred partly into heating which leads to a positive temperature gradient. The heat transport in the lower part of the thermosphere

Table 2.1 Average surface, effective and exobase temperatures of Venus, Earth and Mars

Planets	T_s (K)	T_{eff} (K)	T_{exo} (K)
Venus	~735	~232	~270
Earth	~232	~254	~900
Mars	~215	~210	~220

is dominated by convection, while in the upper part heat transport by conduction takes over, which leads to an isothermal region. The altitude in this region where the mean free part of the atmospheric species becomes equal to their scale height is called the exobase level, which separates the upper atmosphere in a collision-dominated and a collision-less region. The exobase temperature T_{exo} can be analytically be written as [5]

$$T_{exo}^s = \frac{s \, \eta \, \alpha \, k}{K_0 \, m_i \, g} \int_{\lambda_1}^{\lambda_2} d\lambda \, I_{EUV}(\lambda)$$

$$\left\{ E\left[\tau(\lambda)\right] + \ln\left[\tau(\lambda)\right] + \gamma - \frac{m_i}{m_j}\left[1 - e^{-\tau(\lambda)}\right]\right\} + T_0^s, \qquad (2.2)$$

where λ is the wavelength, I_{EUV} is the EUV intensity at the orbit location, τ is the optical thickness, E is the exponential integral, γ is Euler's constant, η is the EUV heating efficiency of the gas, k is the Boltzmann constant, m_i and m_j are the masses of the main atmospheric constituents, $K(T) = K_0 T^s$ is the thermal conductivity coefficient, s depends on the thermospheric composition, g is the gravitational acceleration, and T_0 is the temperature at the base of the thermosphere which is $\approx T_{eff}$. In the case of a hydrogen-dominated thermosphere $i \approx j$ [5]. The factor α is connected to the rotation of the planet and is ~0.25 for rapidly rotating planets and ~0.5 for slowly rotating or tidally locked planets [5, 6]. With some approximations, T_{exo} can be expressed in a less complex formula [6–9]:

$$T_{exo}^s \approx \frac{\eta \, \alpha \, I_{EUV} \, k \, \sigma_c}{K_0 \, m_i \, g \, \sigma_a} + T_0^s, \qquad (2.3)$$

where σ_c and σ_a are the collision and absorption cross-sections. Table 2.1 compares the average surface temperature T_s, the average exobase temperature T_{exo} with the average T_{eff} of Venus, the Earth and Mars. One can see from Table 2.1 that due to the extreme greenhouse effect in Venus' atmosphere the planet's surface temperature is much larger compared to that of Earth and Mars, while the effective temperatures of all three planets are not so different. One can also see that the N_2-dominated atmosphere of the present Earth has a much higher exobase temperature compared to the CO_2-rich upper atmospheres of Venus and Mars. Contrary to nitrogen, CO_2 is a greenhouse gas and also an efficient IR-cooler in the upper atmosphere.

Table 2.2 The evolution of the solar luminosity L, effective temperature T_{eff}, and radius r from the ZAMS until the beginning of the Sun's red giant phase normalized to the present day values [11]

Solar age (Gyr)	L/L_{Sun}	T_{eff}/T_{Sun}	r/r_{Sun}
0	~0.7	~0.9	~0.85
~2	~0.8	~0.95	~0.9
~4	~0.9	~0.97	~0.93
~4.5	1	1	1
~6	~1.05	~1	~1.05
~8	~1.3	~1	~1.15
~10	~1.95	~0.98	~1.5

2.2 Radiation Environment of the Young Sun/Stars

The nuclear evolution of the Sun is well known from stellar evolutionary theory and backed by helioseismological observations of the internal solar structure [10]. The results of these evolutionary solar models, indicate that the young Sun was ~10 % cooler and ~15 % smaller compared to the modern Sun ~4.6 Gyr ago. According to the solar standard model, due to accelerating nuclear reactions in the Sun's core, the Sun is a slowly evolving variable G-type star that has undergone an ~30 % increase in luminosity over the past ~4.5 Gyr. In ~0.1 Gyr from today, the Sun will be ~10 % brighter, so that the Earth will be heated enough that the oceans start to evaporate. In ~6 Gyr from now the Sun expands beyond the Earth's orbit to become a red giant [11, 12]. The evolution of the Sun's luminosity L, effective temperature, T_{eff}, and radius r after it arrived at the ZAMS to the beginning of the red giant phase are shown in Table 2.2. According to these values the onset of ocean evaporation begins at a solar age of ~5.8 Gyr and the runaway greenhouse effect occurs at ~8.5 Gyr. From these parameters the young Sun may have had an initial luminosity of ~70 % of the present Sun resulting in a significantly lower solar constant.

Table 2.3 lists the short wavelength radiation fluxes of the Sun as reconstructed from observed solar proxies as a function of age between ~0.1 and 4.6 Gyr. The fluxes are divided into six wavelength intervals and normalized to the values of the modern Sun [4, 13, 14]. One can see from Table 2.3 that the sample of well-studied solar proxies contain six nearby G0V-G5V stars. The solar analogs EK Dra, π^1 UMa, HN Peg, χ^1 Ori, BE Cet, and κ^1 Cet have known rotation periods, temperatures, luminosities, and metallicities which fall into the early age domain of the Sun to $t \leq 0.65$ Gyr [4, 13].

The observational data of these solar proxies with ages <0.1 Gyr reveal that the EUV flux is saturated at ~100 times of the average present time solar value during the first ~100 Myr [15]. After that very active period the EUV flux decreases with time t in units of Gyr according to an empirical scaling law which can be written as [13, 14]

$$S_{EUV} = \left(\frac{t_{[Gyr]}}{t_{Sun[Gyr]}} \right)^{-1.23}. \tag{2.4}$$

Table 2.3 Solar/stellar radiation flux enhancement factors obtained from observations of solar proxies as function of wavelength normalized to the present solar value from today to ~4.5 Gyr ago [4, 13, 14]

Solar age (Gyr)	t b.p. (Gyr)	X-ray (0.1–2 nm)	SXR (2–10 nm)	EUV (10–92 nm)	FUV (92–118 nm)	Lyman-α (120–130 nm)
4.6	0	1	1	1	1	1
3.2	1.4	2	1.6	1.5	1.4	1.3
2.6	2	3	2	1.9	1.6	1.5
1.9	2.7	6	3	2.7	2.1	1.9
1.1	3.5	16	6	5.1	3.4	2.8
0.7	3.9	37	11	8.6	5	3.9
0.65	3.95	43	12	9.4	5.3	4.1
0.6	4.0	50	13	10	5.7	4.3
0.55	4.05	59	15	11	6.1	4.6
0.5	4.1	71	17	13	6.6	4.9
0.45	4.15	87	19	14	7.2	5.3
0.4	4.2	109	22	17	8	5.8
0.35	4.25	141	26	19	9	6.4
0.3	4.3	189	32	23	10	7.1
0.25	4.35	268	40	28	12	8.1
0.2	4.4	412	54	37	14	9.6
0.15	4.45	715	77	51	18	11.8
0.1	4.5	1558	129	82	26	15.8

with the EUV flux enhancement factor S_{EUV} and t_{Sun} the age of the Sun in units of Gyr. It should be noted that by expanding the activity-age relation from solar mass to lower mass stars one finds that lower mass stars spend more time in the saturated level before their activity decays [16] and stars of the same spectral type show a wider distribution of the X-ray luminosity [17]. By using ROSAT satellite measurements of G-type stars which cover the transition region in wavelength λ between SXR and EUV ($\lambda \sim 10$ nm) one can construct scaling laws based from the ROSAT data which reproduce the EUV and SXR luminosity evolution as a valid EUV proxy [17–20]. These scalings are justified because the SXR data correlate well with measurements of the EUVE satellite in the wavelength range of ~10–36 nm shown for the solar proxies in Table 2.2. From this relation one can also construct generalized EUV scaling laws for F, G, K, and M-type stars for L_{SXR} in units of erg s^{-1}. The temporal evolution of the stellar SXR flux at a planetary orbit and r_{pl} are taken into account, while the stellar mass and orbital distance d are assumed to be constant. For the range of general F stars with $L_0 = 10^{29.83}$ one obtains [19],

$$L_{SXR} = 0.284 L_0 t^{-0.547} t \leq 0.6 \, \text{Gyr},$$
$$L_{SXR} = 0.155 L_0 t^{-1.72} t > 0.6 \, \text{Gyr}, \tag{2.5}$$

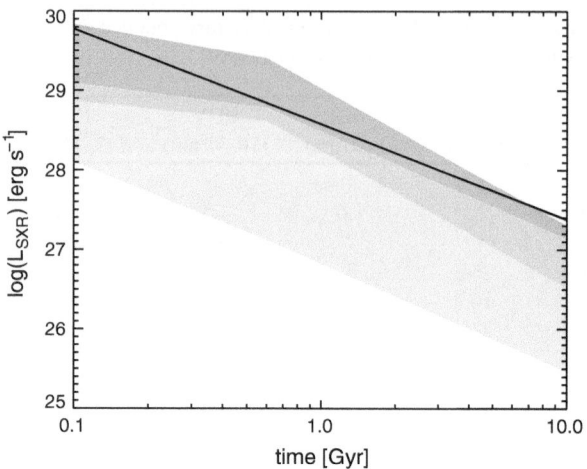

Fig. 2.2 Average SXR luminosity L_{SXR} in units of erg s^{-1} as a function of stellar age for G- and F-type stars (*dark gray area*), K and M stars (*bright gray area*) [19] at 1 AU. The *solid line* represents the result from the power law analyzed from EUVE satellite data of solar proxies with different ages [13]. The *dark gray* together with the *gray area* correspond to the standard deviation 1 σ of the G-star distribution, and *light gray* together with the *gray area* to the standard deviation 1 σ of M-type stars. Both distributions are overlapped by the *gray area*

for G-stars with $L_0 = 10^{29.35}$,

$$L_{SXR} = 0.375 L_0 t^{-0.425} t \leq 0.6 \, \text{Gyr},$$
$$L_{SXR} = 0.19 L_0 t^{-1.69} t > 0.6 \, \text{Gyr}, \tag{2.6}$$

for K-type stars, $L_0 = 10^{28.87}$,

$$L_{SXR} = 0.474 L_0 t^{-0.324} t \leq 0.6 \, \text{Gyr},$$
$$L_{SXR} = 0.234 L_0 t^{-1.72} t > 0.6 \, \text{Gyr}, \tag{2.7}$$

and for M dwarfs, $L_0 = 10^{28.75}$,

$$L_{SXR} = 0.17 L_0 t^{-0.77} t \leq 0.6 \, \text{Gyr},$$
$$0.13 L_0 t^{-1.34} t > 0.6 \, \text{Gyr}. \tag{2.8}$$

The stellar age t is in units of Gyr and $F_{EUV} \approx L_{SXR}/4\pi d^2$ [19]. Figure 2.2 shows the SXR luminosity evolution obtained from the above-mentioned scaling laws for G, F, K, and M-type dwarf stars as a function of stellar age at an orbital distance of 1 AU. One can see that stars with lower masses remain somewhat longer at active emissions compared to solar-like G-type stars. After their high activity phase the emissions decrease by a similar power law relationship as that of the solar proxies shown in

Table 2.3. In addition to the short wavelength radiation, which can heat and expand upper atmospheres the solar/stellar wind, as well as coronal mass ejections (CMEs) will also result in permanent forcing of the thermosphere-exosphere-magnetosphere environment during the early phase of a planet's lifetime.

2.3 Evolution of the Solar/Stellar Wind with Time

Because the outward flowing plasma carries away angular momentum from the star one of the best indirect confirmation that the plasma outflow from young stars is very efficient is the observed spin-down to slower rotation of young stars after their arrival at the ZAMS. Stellar wind and CME-related plasma are ejected from the star but the observation of such features at solar proxies is very difficult [4, 21–23]. Detection methods of stellar winds include the measurement of thermal radio emission from the plasma flow [24, 25], and signatures of charge exchange in X-ray spectra [26].

The most successful approach so far is an indirect method which observes the Lyman-α absorption in astrospheres caused by neutral hydrogen atoms which are produced from the interaction between stellar wind plasma and the interstellar medium (ISM) [22, 23]. A termination shock forms when stellar winds collide with the ISM at further distance from the star and the wind is shocked to subsonic velocities [21–23]. Much of the gas is piled up between the heliopause and the bow shock. As a result a hydrogen wall is produced and its signature in the Lyman-α line of neutral H atoms can be observed. The measurable absorption depths in the Lyman-α line are then compared with results from hydrodynamic model calculations [23].

A systematic study of all derived stellar mass-loss rates of main sequence stars which were observed with this method indicates that the mass loss as a function of time dM_L/dt per unit stellar surface correlates well with the stellar activity and the stellar X-ray surface flux, $F_{X\text{-ray}}$ [23]

$$\frac{dM_L}{dt} \propto F_{X\text{-ray}}^{1.34\pm0.18}. \tag{2.9}$$

When one extrapolates the above power law up to the observed X-ray saturation limit of the Sun-like stars of $\sim2 \times 10^7\,\text{erg}\,\text{cm}^{-2}\,\text{s}^{-1}$, one would suggest that the average mass loss of plasma from the youngest stars such as these shown in Table 2.3 is more than $\sim$1,000 times larger compared to the mass loss of the present Sun. However, until now there are no accurate mass loss observations of young solar-like single stars available. To avoid wrong estimations the power law relation given in Eq. 2.9 should not be applied for the most active stars for which $F_X \geq 8 \times 10^5\,\text{erg}\,\text{cm}^{-2}\,\text{s}^{-1}$ [23].

A possible reason that the mass loss-X-ray activity power law relation fails for very young solar-type stars could be related to the fast rotation period of young stars of a few days during the first 500 Myr after the arrival at the ZAMS because the activity and magnetic field configuration was most likely quite different from the

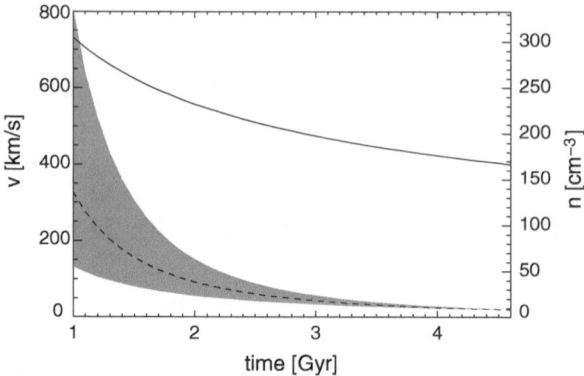

Fig. 2.3 *Dashed-line* average stellar wind density and velocity (*solid line*) estimated from stellar mass loss observations at 1 AU between 1 Gyr and present time (4.6 Gyr)

current appearance [27, 28]. In such a case the Sun spots at latitudes which are $\geq 60°$ degree could change the structure of coronal magnetic fields as well as the formation of prominences, which might lead to a predominant occurrence of flares and CMEs at high latitudes which may propagate significantly above the ecliptical plane. For studying the effect on planetary environments against solar/stellar wind-induced non-thermal escape, the solar/stellar wind density and velocity in the vicinity of a planet are key parameters. The age dependence of the solar/stellar wind velocity for the Sun-like stars can be estimated from the stellar mass loss observations discussed above, in combination with an empirical models [4, 29–31]. The shaded area in Fig. 2.3 represents uncertainties of the solar wind density in time which shows the expected enhancement in solar wind density and velocity, according to mass loss observations of the Sun-like stars [23] and the power laws [31] as function of age from present time back to ∼3.5 Gyr ago. From the observational evidence of the stellar radiation and plasma environment one can expect that these photon and particle fluxes certainly have an impact on the origin and early evolution of planetary atmospheres.

References

1. Lundin, R., Lammer, H., Ribas, I.: Planetary magnetic fields and solar forcing: implications for atmospheric evolution. Space Sci. Rev. **129**, 245–278 (2007)
2. Sagan, C., Mullen, G.: Earth and Mars: evolution of atmospheres and surface temperatures. Science **177**, 52–56 (1972)
3. Krauss, S., Fichtinger, B., Lammer, H., Hausleitner, W., Kulikov, Yu. N., Ribas, I., Shematovich, V.I., Bisikalo, D., Lichtenegger, H.I.M., Zaqarashvili, T.V., Khodachenko, M.L., Hanslmeier, A.: Solar flares as proxy for the young Sun: satellite observed thermosphere response to an X17.2 flare of Earth's upper atmosphere. Ann. Geophys. **30**, 1129–1141 (2012)
4. Lammer, H., Güdel, M.: Kulikov, YuN, Ribas, I., Zaqarashvili, T.V., Khodachenko, M.L., Kislyakova, K.G., Gröller, H., Odert, P., Leitzinger, M., Fichtinger, B., Krauss, S., Hausleitner,

W., Holmström, M., Sanz-Forcada, J., Lichtenegger, H.I.M., Hanslmeier, A., Shematovich, V.I., Bisikalo, D., Rauer, H., Fridlund, M.: Variability of solar/stellar activity and magnetic field and its influence on planetary atmosphere evolution. Earth Planets Space **63**, 179–199 (2012)

5. Gross, S.H.: On the exospheric temperature of hydrogen-dominated planetary atmospheres. J. Atmos. Sci. **29**, 214–218 (1972)

6. Bauer, S.J.: Physics of Planetary Ionospheres. Springer, Berlin (1973)

7. Bauer, S.J.: Solar cycle variation of planetary exospheric temperatures. Nature **232**, 101–102 (1971)

8. Lammer, H., Selsis, F., Ribas, I., Guinan, E.F., Bauer, S.J., Weiss, W.W.: Atmospheric loss of exoplanets resulting from stellar X-Ray and extreme-ultraviolet heating. ApJ **598**, L121–L124 (2003)

9. Bauer, S.J., Lammer, H.: Planetary Aronomy: Atmosphere Environments in Planetary Systems. Springer, Berlin (2004)

10. Guzik, J.A., Watson, L.S., Cox, A.N.: Implications of revised solar abundances for helioseismology. Memorie della Societa Astronomica Italiana **77**, 389–392 (2006)

11. Bressan, A., Fagotto, F., Bertelli, G., Chiosi, C.: Evolutionary sequences of stellar models with new radiative opacities. II - Z=0.02. Astron. Astrophys. Suppl. Ser. **100**, 647–664 (1993)

12. Kasting, J.F.: Runaway and moist greenhouse atmospheres and the evolution of Earth and Venus. Icarus **74**, 472–494 (1988)

13. Ribas, I., Guinan, E.F., Güdel, M., Audard, M.: Evolution of the solar activity over time and effects on planetary atmospheres. I. High-energy irradiances (1–1700Å). ApJ **622**, 680–694 (2005)

14. Güdel, M.: The sun in time: activity and environment. Living Rev. Sol. Phys. **4**(3), 1–137 (2007)

15. Güdel, M., Guinan, E.F., Skinner, S.L.: The X-ray sun in time: a study of the long-term evolution of coronae of solar-type stars. ApJ **483**, 947–960 (1997)

16. Scalo, J., Kaltenegger, L., Segura, A.G., Fridlund, M., Ribas, I.: Kulikov, Yu.N., Grenfell, J.L., Rauer, H., Odert, P., Leitzinger, M., Selsis, F., Khodachenko, M.L., Eiroa, C., Kasting, J., Lammer, H.: M stars as targets for terrestrial exoplanet searches and biosignature detection. Astrobiology **7**, 85–166 (2007)

17. Penz, T., Micela, G., Lammer, H.: Influence of the evolving stellar X-ray luminosity distribution on exoplanetary mass loss. Astron. Astrophys. **477**, 309–314 (2008)

18. Penz, T., Micela, G.: X-ray induced mass loss effects on exoplanets orbiting dM stars. Astron. Astrophys. **479**, 579–584 (2008)

19. Lammer, H., Odert, P., Leitzinger, M., Khodachenko, M.L., Panchenko, M., Kulikov, YuN, Zhang, T.L., Lichtenegger, H.I.M., Erkaev, N.V., Wuchterl, G., Micela, G., Penz, A., Biernat, H.K., Weingrill, J., Steller, M., Ottacher, H., Hasiba, J., Hanslmeier, A.: Determining the mass loss limit for close-in exoplanets: what can we learn from transit observations? A&A **506**, 399–410 (2009)

20. Lammer, H., Bredehöft, J.H., Coustenis, A., Khodachenko, M.L., Kaltenegger, L., Grasset, O., Prieur, D., Raulin, F., Ehrenfreund, P., Yamauchi, M., Wahlund, J.-E., Grießmeier, J.-M., Stangl, G., Cockell, C.S., Kulikov, Yu.N., Grenfell, L., Rauer, H.: What makes a planet habitable? Astron. Astrophs. Rev. **17**, 181–249 (2009)

21. Wood, B.E., Müller, H.-R., Zank, G., Linsky, J.L.: Measured mass loss rates of solar-like stars as a function of age and activity. ApJ **574**, 412–425 (2002)

22. Wood, B.E.: Astrospheres and solar-like stellar winds. Living Rev. Sol. Phys. **1**(2), 1–44 (2004)

23. Wood, B.E., Müller, H.-R., Zank, G.P., Linsky, J.L., Redfield, S.: New mass loss measurements from astrospheric Ly-α absorption. ApJ **628**, L143–L146 (2005)

24. Lim, J., White, S.M.: Limits to mass outflows from late-type dwarf stars. ApJL **462**, L91–L94 (1996)

25. Gaidos, E.J., Güdel, M., Blake, G.A.: The faint young Sun paradox: an observational test of an alternative solar model. Geophys. Res. Lett. **27**, 501–503 (2000)

26. Wargelin, B.J., Drake, J.J.: Observability of stellar winds from late-type dwarfs via charge exchange X-ray emission. ApJL **546**, L57–L60 (2001)
27. Zaqarashvili, T.V., Oliver, R., Ballester, J.L., Carbonell, M., Khodachenko, M.L., Lammer, H., Leitzinger, M., Odert, P.: Rossby waves and polar spots in rapidly rotating stars: implications for stellar wind evolution. A and A **532**, A139 (2011)
28. Zaqarashvili, T.V., Carbonell, M., Oliver, R., Ballester, J.L.: Quasi-biennial oscillations in the solar tachocline caused by magnetic Rossby wave instabilities. ApJL **724**, L95–L98 (2011)
29. Grießmeier, J.M., Motschmann, U., Stadelmann, A., Penz, T., Lammer, H., Selsis, F., Ribas, I., Guinan, E.F., Biernat, H.K., Weiss, W.W.: The effect of tidal locking on the magnetospheric and atmospheric evolution of "Hot Jupiter". Astron. Astrophys. **425**, 618–630 (2007)
30. Grießmeier, J.-M., Preusse, S., Khodachenko, M., Motschmann, U.: Exoplanetary radio emission under different stellar wind conditions. Planet. Space Sci. **55**, 618–630 (2007)
31. Newkirk Jr., G.: Solar variability on time scales of 10^5 years to $10^{9.6}$ years. Geochi. Cosmochi. Acta Suppl. **13**, 293–301 (1980)

Chapter 3
Escape of Planetary Atmospheres

The extreme radiation and plasma environments during the period of the young active Sun/Stars have important implications for the evolution of planetary atmospheres and may be responsible that planets with a low gravity like early Mars most likely could never build up a dense atmosphere during the first few 100 Myr after their origin. On the other hand more massive planets such as super-Earths even in orbits within the habitable zone of their host stars might not lose their initial protoatmospheres completely. These planets could end up as water worlds [1, 2] with CO_2 and hydrogen- or O-rich upper atmospheres. If an atmosphere of a terrestrial planet evolves to an N_2-dominated atmosphere too early in its lifetime, the atmosphere may escape to space. By comparing the escape-related atmospheric evolution between Venus, the Earth and Mars, one finds that the initial conditions set up by the planetary formation processes and the interaction between the early atmospheres with the young Sun's or host star's X-ray and EUV flux as well as the plasma environment (e.g., winds, CMEs, etc.) influence strongly the factors to which a planet may evolve to an Earth-like class I habitat.

3.1 Atmospheric Escape Processes and their Role in Atmosphere Evolution

Satellite observations and theoretical studies have shown that enhanced solar/stellar EUV radiation and plasma flows (e.g., winds, CMEs) result in a continuous forcing of the upper region of planetary atmospheres (Fig. 3.1), which can ionize, heat, expand, chemically modify, and erode it during the early phase of a planetary lifetime (Fig. 3.2) [3–11]. On the basis of the sources of the host stars energy input into the upper atmosphere, we can separate two main escape categories: thermal escape and non-thermal escape. The thermal escape of atmospheric particles can be separated in two regimes:

H. Lammer, *Origin and Evolution of Planetary Atmospheres*,
SpringerBriefs in Astronomy, DOI: 10.1007/978-3-642-32087-3_3,
© The Author(s) 2013

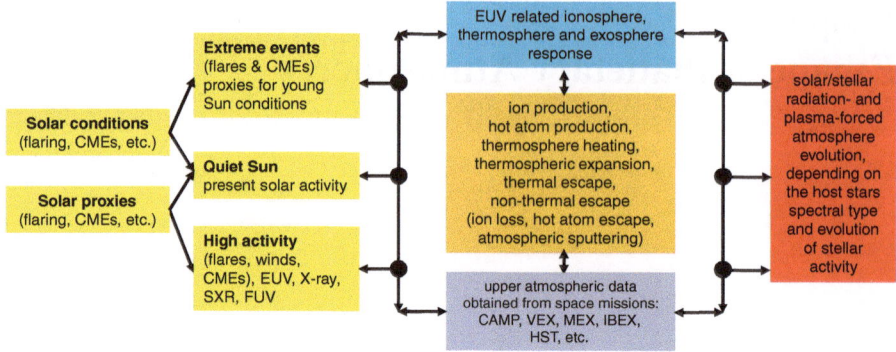

Fig. 3.1 Illustration of the induced response of upper atmospheres related to the radiation and plasma forcing of the young Sun or active stars

- *Jeans escape*: evaporation of particles which populate the high energy tail of a Maxwell distribution,

and

- *hydrodynamic blow-off*: evaporation of the whole exosphere.

When the temperature of the thermosphere is high, a significant part of the lighter constituents (H, D, H_2, He) of the upper atmosphere obtain velocities above the escape value, and the hydrostatic thermosphere can change to a non-hydrostatic condition (see Figs. 3.2 and 3.3). In such a case the thermosphere starts accompanied by adiabatic cooling dynamically to expand to several planetary radii [4, 6, 10–15].

Depending on the mass and size of a planet and the stellar EUV flux value, in some cases the thermosphere can even expand beyond an atmosphere protecting magnetopause [16], so that non-thermal escape processes as illustrated in Fig. 3.4 become relevant. Most non-thermal atmospheric loss processes such as

- *ion pick up*: planetary atoms are ionized and accelerated by electric fields within the solar/stellar wind plasma flow around a planetary obstacle,
- *plasma instabilities*: wave structures at a plasma boundary (i.e., ionopause) can detach ionospheric clouds,
- *momentum transport and cool ion outflow*: planetary ions can be accelerated by the solar/stellar wind plasma flow as well as by electrical fields to escape velocities throughout the tail of non- or weakly magnetized planets,
- *polar wind*: ion outflow over polar regions or magnetospheric cusps of magnetized planets,

are related to the escape of ionized atmospheric particles. Only photochemical processes such as

- *dissociative recombination*: $A_2^+ + e \rightarrow A^* + A^* + \Delta E$,
- *photodissociation*: $A_2 + \nu \rightarrow A^* + A^* + \Delta E$,

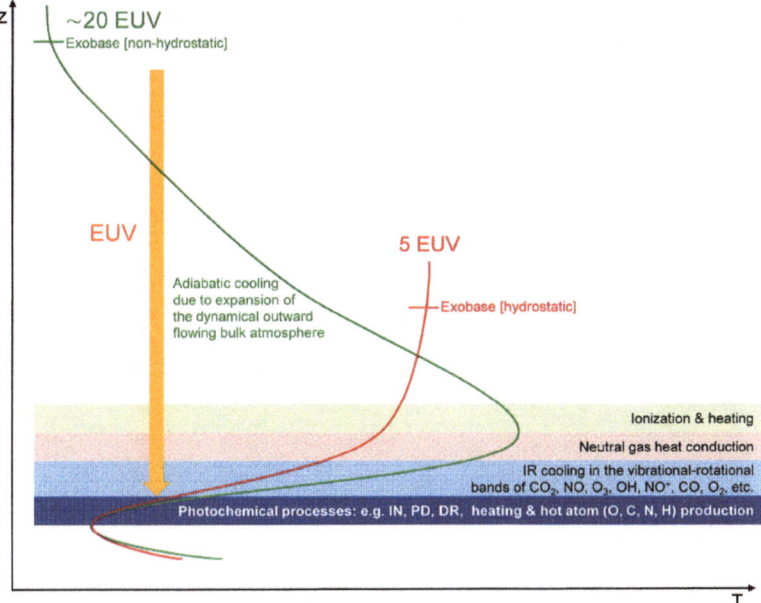

Fig. 3.2 Illustration of the response of the thermospheric temperature structure to a high solar/stellar EUV flux. Enhanced ionization and photochemical processes eventually lead to heating and subsequent expansion of the upper atmosphere and to the production of suprathermal atoms which can also influence the energy balance in the thermosphere. Depending on the atmospheric species and the deposited energy, upper atmospheres can change from hydrostatic [6, 9] to hydrodynamic regimes [10, 11]

- *charge exchange*: $A^+ + B^* \rightarrow A + B^+$,

and

- *atmospheric sputtering*: $A^+ + B \rightarrow A^+ + B^*$,

produce excited neutral atoms, which form extended planetary coronae or suprathermal atom populations within a planet's exosphere. The released energy $\Delta E = E_I - E_D - E(*)$, where E_I equals the ionization energy, E_D the dissociation energy, and $E(*)$ denotes the internal excitation energies for the corresponding atomic states. In the case of low mass planetary bodies such as Mars atoms which can overcome the escape velocity.

Highly irradiated terrestrial planets may thus be even in danger of being stripped of their whole atmospheres [7, 16–18]. On the other hand if the thermal and non-thermal atmospheric escape processes are too weak, a planet may have problems to get rid of its protoatmosphere which will have essential implications for habitability.

In the following sections, all known atmospheric escape processes which can act on neutral and ionized particles will be briefly discussed. It will be shown that the stellar EUV fluxes as well as the strength of the plasma outflow form a star control the efficiency of all atmospheric escape processes and play a relevant role during the planet's atmospheric evolution stage. Thus, thermal and non-thermal atmospheric escape processes are closely connected.

3.1.1 Jeans Escape

In classical Jeans escape, particles at the exobase level are assumed to have a Maxwell velocity distribution which is determined by the temperature at the exobase level, z_{exo}. This level separates the collision-dominated region from the exosphere where collisions become negligible. The exobase distance in any planetary atmosphere can be estimated by

$$\int_{z_{exo}}^{\infty} \frac{dz}{l(z)} = \int_{z_{exo}}^{\infty} n(z)\sigma_c \, dz = n_c H \sigma_c \approx \frac{H}{l} = 1, \tag{3.1}$$

with the mean free path $l = (n\sigma_c)^{-1}$, the collision cross-section σ_c, and atmospheric scale height $H = (kT_{exo})/gm$ at the exobase, where k is the Boltzmann constant, T_{exo} the atmospheric temperature at the exobase, g the gravitational acceleration and m the mass of the atmosphere species. Classical thermal escape therefore occurs at atmospheric altitudes where the mean free path $l \sim H$. The exobase density can thus be estimated by

$$n_{exo} = (\sigma_c H_{exo})^{-1}, \tag{3.2}$$

with H_{exo} the scale height at the exobase level. At that altitude, particles of the high velocity tail of the Maxwellian velocity distribution which move upward with velocities $v \geq v_{esc}$

$$v_{esc} = \left(\frac{2GM_{pl}}{r}\right)^{\frac{1}{2}} = \sqrt{2gr}, \tag{3.3}$$

so that the particles overcome the gravitational potential of the planet and will be lost from the upper atmosphere [19]. G is the gravitational constant, M_{pl} is the mass of the planet and r is the radial distance from the planetary center. One should note that a sharply defined exobase distance represents only a highly idealized concept and in reality the exobase can be considered as a region where collision processes become negligible. The escape efficiency depend on how frequent or infrequent collisions between the fast moving atoms with the slower background atmospheric particles occur. The probability for an atom or molecule traveling a distance z without a collision is given by the collision probability

$$P(z) = \exp\left(-\frac{z}{l}\right) \tag{3.4}$$

where the mean free path l represents the distance for which the collision probability is e^{-1}. As long as the upper atmosphere remains under hydrostatic conditions one can assume that below and close to the exobase a Maxwell velocity distribution prevails. In the exosphere, the velocity distribution is truncated due to absence of collisions and particles with velocities above v_{esc}. Therefore, the barometric law applies only in the collision- dominated atmosphere below the exobase. Thus, the basic assumptions of Jeans escape can be summarized as

- an isothermal atmosphere, and
- a Maxwell velocity distribution even at the escape level.

The outward flux of atmospheric particles with $v \geq v_{esc}$ at a planetocentric distance r results in the Jeans escape flux F_{Jeans}, which is obtained by the product of the Maxwell velocity distribution function and the vertical velocity component ($v > v_{esc}$) integrated over the hemisphere at the exobase distance. The Jeans escape flux can thus be written as [19–21]

$$F_{Jeans} = \frac{v_{max}}{2\sqrt{\pi}} \cdot n_{exo}(1 + X_{exo})e^{-X_{exo}}, \qquad (3.5)$$

with n_{exo} is the density of the escaping constituent at the exobase and the most probable velocity at a Maxwellian velocity distribution $v_{max} = (2kT_{exo}/m)^{\frac{1}{2}}$ at the exobase. The so-called escape parameter X_{exo} at the exobase level $r \to r_{exo}$ can be generally expressed by [21]

$$X(r) = \frac{GM_{pl}m}{rkT_{esc}} = \frac{v_{esc}^2}{v_0^2}. \qquad (3.6)$$

On planets which rotate fast, atoms with velocity trajectories in the direction of the rotational motion can easier reach v_{esc} than particles which move in the opposite direction [20]. As a result the flux of the escaping particles will therefore be directed in the forward direction.

One should also note that the escaping atmospheric particles itself lead to a perturbation of the Maxwell velocity distribution which can result in an overestimation of the Jeans escape rate. Statistical Monte Carlo treatments indicate that the classical Jeans escape formula overestimates the escape flux in the worst cases by $\sim 30\%$ [21–23].

Jeans escape becomes relevant if the escape parameter $X_{exo} < 15$. An atmosphere is gravitationally bound to a planet for values of $X \geq 30$, while for $X_{exo} \leq 1.5$ the upper atmosphere becomes unstable and the whole exosphere evaporates [24]. In classical Jeans escape, the velocity of the evaporating particles is always subsonic. In a case if all atoms of the exosphere exceed their gravitational potential energy of the planet, thermal escape can be considered as hydrodynamic blow-off.

3.1.2 Hydrodynamic Atmospheric Expansion
    and Blow-Off

Planetary thermospheres can mainly be classified into two regimes. In the first regime the upper atmosphere is in hydrostatic equilibrium. In such condition, the bulk of the thermosphere below the exobase level can be considered as static. In the second regime, the upper atmosphere expands hydrodynamically in which the bulk atmospheric particles in the thermosphere can escape efficiently as a result of high solar- or stellar EUV fluxes, energy deposited by particles, and/or a weak planetary gravitational field. Under certain EUV conditions and planetary parameters the upper atmosphere can hydrodynamically expand but not all atoms may reach escape velocity [10, 11]. In such a case one can expect a hydrodynamically expanding thermosphere where the bulk gas does not reach escape velocity at the exobase level and the loss results in a strong Jeans-type escape but not in a hydrodynamic blow-off.

If the mean thermal energy of the upper atmosphere gases at the exobase level exceed their gravitational energy, blow-off occurs. Hydrodynamic blow-off can be considered as the most efficient atmospheric escape process. In this extreme condition, the atmospheric escape is very high because the whole exosphere evaporates and will be refilled by the upward flowing planetary gas of the dynamically expanding thermosphere as long as the thermosphere can remain in this extreme condition.

For studying the atmospheric structure of an hydrodynamically expanding upper atmosphere one can apply the set of hydrodynamic equations for mass, momentum, and energy conservation in spherical coordinates

$$
\frac{\partial n}{\partial t} + \frac{1}{r^2} \frac{\partial n v r^2}{\partial r} = 0,
$$
$$
n \frac{\partial v}{\partial t} + n v \frac{\partial v}{\partial r} + \frac{1}{m} \frac{\partial p}{\partial r} = n F_{\text{grav}}, \tag{3.7}
$$
$$
n m \left(\frac{\partial E}{\partial t} + v \frac{\partial E}{\partial r} \right) = q - p \frac{1}{r^2} \frac{\partial r^2 v}{\partial r} + \frac{1}{r^2} \frac{\partial}{\partial r} \left(r^2 \chi \frac{\partial T}{\partial r} \right),
$$

where

$$
p = nkT, \quad E = \frac{1}{\gamma - 1} \frac{p}{n m}. \tag{3.8}
$$

Here, n corresponds to the atmospheric number density, v is the velocity of the dynamically outward moving atmospheric particles, m is the mass of the atmospheric particles, p is the thermal gas pressure, E is the total energy density of the atmospheric gas, q is the solar or stellar EUV volume heating rate, T is the atmospheric gas temperature, k is Boltzmann's constant, γ is the adiabatic index of the atmospheric gas, and χ is the heat conductance. For exoplanets which are orbiting very close around their host stars, one cannot neglect gravitational effects related to the Roche lobe [25] by considering

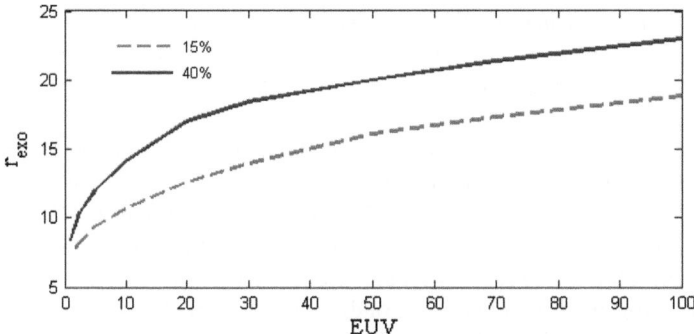

Fig. 3.3 Hydrodynamical expanded exobase location in units of planetary radii of an atomic H populated thermosphere of an Earth-size and mass planet as a function of solar/stellar EUV flux for a heating efficiency of 15 % (*dashed-line*) and of 40 % (*solid-line*) (courtesy of N. V. Erkaev)

$$F_{grav} = -\frac{GM_{pl}}{r^2} + \frac{GM_{st}}{(d-r)^2} - \frac{G(M_{st} - M_{pl})}{d^3}(s-r), \qquad (3.9)$$

where G is the gravitational constant, M_{pl} is the planetary mass, M_{st} is the stellar mass, d is the orbital distance of the planet, and s is distance of the center of mass of the system from the planet's center. By comparing the thermal and kinetic energies during high EUV fluxes one finds that a substantial amount of the heating goes into the acceleration and thermal energy of the exposed hydrogen atoms. Generally, the total absorbed energy is distributed into the following three parts

- in the energy flux of the escaping particles,
- in the thermal energy of the escaping particles,
- in the kinetic energy of the escaping particles.

Figure 3.3 shows the application of a hydrodynamic model which is based on the model described in detail in [25] to an Earth-size and mass planet with an atomic hydrogen-rich thermosphere in the habitable zone which is exposed to an EUV flux which is 1–100 times larger compared to that of the present Sun. The temperature at that base of the thermosphere is assumed to be $\sim$250 K [15] and the atmospheric H atom number density is assumed to be in the order of $\sim$5 $\times$ 10^{12} cm^{-3}. In agreement with [10, 15] under such conditions a hydrogen-dominated upper atmosphere is not hydrostatic but changes into a hydrodynamic regime.

One can see in Fig. 3.3 that the exobase level expands even for a low heating efficiency, which can be described as the ratio of the net heating rate to the rate of stellar energy absorption, of 15 % to a distance which is located at $\sim$7.5 for 1 EUV, $\sim$9.5 for 5 EUV, $\sim$15 for 40 EUV and $\sim$17.3 for 70 EUV and $\sim$19 Earth-radii for 100 EUV. If one chooses a larger heating efficiency of 40 % the expansion of the exobase level reaches altitude locations which are $\geq$20 Earth-radii for EUV flux values which are >50 times that of today's Sun. If there are no additional IR-cooling molecules such as CO_2 or H_3^+, the whole thermosphere changes for a heating efficiency of 15 %

in the blow-off state when the EUV flux is ≥ 10 times of today's Sun. For a heating efficiency of 40 % blow-off can even be reached for EUV flux values ≥ 5 times that of the present Sun. In such a case the H atom escape rates reaches values which are $\geq 10^{30}$ s^{-1}.

In very early extreme environments soon after a protoatmosphere is produced and the atmosphere if hit by many impactors, the lower thermosphere can be much hotter than 250 K and even reach $\sim 1,500$ K. In such an environment together with high X-ray and EUV fluxes the atmosphere could escape more or less energy-limited, which means that ~ 100 % of the incoming energy powers the escape of the atmosphere. However, such extreme conditions and high atmospheric escape fluxes may act only a few ten Myr after the planet's origin.

From the results of such studies, together with observations of expanding hydrogen-rich atmospheres on exoplanets [26], one can conclude that hydrodynamic flow and associated adiabatic cooling may have existed in the thermospheres of early terrestrial planets after they obtained their hydrogen- or H_2O-rich protoatmospheres. Thus, a wide range of early and/or close-in terrestrial-type exoplanets will experience thermospheric expansion and related adiabatic cooling during their atmospheric evolution.

3.1.3 Ion Pick Up

Exospheric neutral particles which follow ballistic trajectories can be ionized and picked up by charge exchange with solar- or stellar- wind protons, due to electron impact and EUV radiation. This non-thermal atmospheric escape process as illustrated in Figs. 3.4 and 3.5 is very efficient in case a planet such as Venus or Mars which is not protected by a strong magnetosphere, or in case of planets such as the early Earth when the upper atmosphere was dynamically expanded above the possible magnetopause.

As illustrated in Fig. 3.5, the total production rate of planetary ions A^+ is the sum of the rates of photoionization, electron impact, and charge exchange with neutral atoms "A" from the upper atmosphere of a planet. For studying the ion pick up loss rate one can simulate the particle flux by dividing the space around the planetary obstacle, which could be an ionopause or a magnetopause into a number of volume elements ΔV. The solar or stellar wind proton flux Φ_{sw} in a volume element ΔV_i at position $\mathbf{r}_i$ as a function of the planetary center can then be written as

$$\Phi_{sw}(\mathbf{r}_i) = \Phi_{sw}^{(0)}(\mathbf{r}_i)e^{-\int\limits_{\infty}^{s_i}\sum\limits_{\alpha} n_\alpha \sigma_\alpha \, ds}, \tag{3.10}$$

where the integration is performed from the upstream solar wind to the corresponding point s_i at position $\mathbf{r}_i$ on the streamline. Here, $\Phi_{sw}^{(0)}$ is the unperturbed solar wind flux, n_A the density of neutral species "A" as a function of altitude, σ_A is the

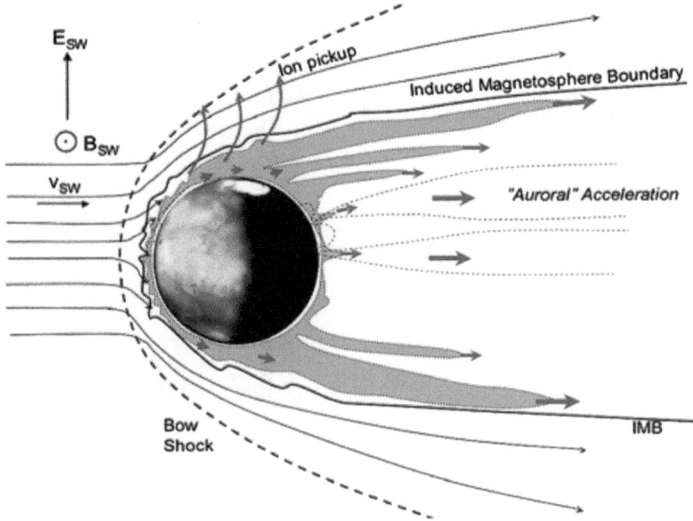

Fig. 3.4 Illustration of direct solar-/stellar-wind plasma forcing on an upper atmosphere of an non- or weakly magnetized terrestrial planet such as Venus or Mars. Ions will be picked up above the planetary obstacle and cold ions can be accelerated by electric fields throughout the tail of the planet. In case of Mars, magnetic anomalies can act similar as at the Earth's polar region in a kind of Auroral ion acceleration (after [7])

energy-dependent charge exchange cross- section between a proton and the neutral constituent "A" and ds is the line element along the streamline. The loss rates of solar wind protons l_{sw} due to the interaction with atmospheric neutral atoms "A" can be written as [27, 28]

$$l_{sw}^{A} = \Phi_{sw} n_A \sigma_A, \tag{3.11}$$

and the corresponding planetary ion production rates p due to charge exchange are simply equal to the corresponding loss rates of the solar wind or the production of energetic neutral atoms (ENAs), i.e., $p_{A^+}^{ce} = l_{sw}^{A^+}$ with A^+ denoting the ion species. The rate of ions produced by electron impact is given by $p_{\alpha}^{ei} = v n_e n_{\alpha}$, where v is the ionization rate coefficient and n_e the electron density. The total planetary ion production rate of species α in each volume element is the sum

$$p_{A^+}^{tot} = p_{A^+}^{ei} + p_{A^+}^{ce} + p_{A^+}^{\gamma}, \tag{3.12}$$

with $p_{A^+}^{\gamma}$ being the rate due to photoionization. To determine the flux of A^+ ions from each volume element, a test particle, which is considered to represent all particles in the volume, is launched and its trajectory followed by integrating the equation of motion [27]

Fig. 3.5 Illustration of the
ion pick up escape process.
After an exospheric neutral
atom reaches the solar/stellar-
wind interaction region, above
a planetary obstacle it can
be an ionopause or a mag-
netopause, it can be ionized
via charge exchange with
a solar/stellar-wind proton
which is transferred to an
energetic neutral hydrogen
atom (ENA), via electron
impact or photon collisions.
The newborn planetary ion
will be picked up from the
exosphere and incorporated in
the surrounding plasma flow

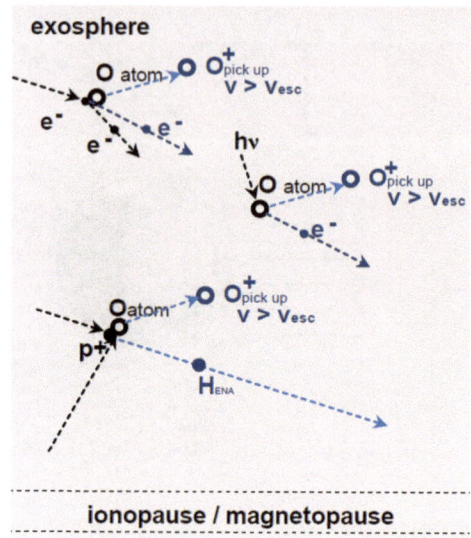

$$\frac{d^2\mathbf{x}}{dt^2} = \frac{q}{m}\left(\mathbf{E} + v \times \mathbf{B}\right), \tag{3.13}$$

where q is the particle charge, m is the particle mass, v is the particle velocity, $\mathbf{B}$ is the
magnetic field, and $\mathbf{E} = -v_{sw} \times \mathbf{B}$ is the motional electric field, i.e., we assume that
the magnetic field is frozen into the flow. The total ion flux associated with particles
of species α through an interaction area IA which is born inside the element $\Delta V^{(i)}$
finally becomes [27]

$$\Phi_{A+}^{(i)} = \frac{p_\alpha^{tot}\Delta V^{(i)}}{IA}, \tag{3.14}$$

where p_{A+}^{tot} has to be taken at the creation point of the particle. The efficiency of
all three ion loss processes depend on the plasma flow proton density, the EUV flux
and the electron density in the plasma, as well as the gyroradius of the picked up
planetary ion. If the gyroradius of the picked up ion is comparable to the planetary
radius a huge fraction of the ions can be backscattered to the upper atmosphere so
that the escape rate will be reduced.

3.1.4 Plasma Instabilities and Ionospheric Clouds

Besides the non-thermal ion pick up escape process, observations of wave-like structures and plasma clouds in the vicinity of Venus by Pioneer Venus Orbiter (PVO) let
to suggestions that plasma instabilities such as the Kelvin–Helmholtz (KH) instability or the interchange instability could be relevant at least around planets with weak

or no intrinsic magnetic fields [29–35]. When the solar or stellar wind plasma flows around an ionosphere the interface between the solar wind and the ionosphere, the ionopause can be a subject to these plasma instabilities. Disturbances of the interface between the flow of two plasma regions with different densities can grow with time, so that the interaction boundary becomes unstable. The velocity shear between two separated plasma regions can lead to the KH-instability [32, 33, 36]. On the other hand the curvature of the magnetic field is the cause of the interchange instability [34].

In the case of the KH-instability, wave-like structures of initially small amplitudes grow and eventually reach a nonlinear stage, so that plasma vortices can develop on their way along the plasma boundary layer from the subsolar/stellar point to the planetary terminator. If a plasma vortex structure develops it might be able to detach ions from the ionized population of the upper atmosphere which escape from the planet in so-called detached plasma clouds [30, 33].

Observations by the Venus Express (VEX) magnetometer VEXMAG were reported, which indicate that vortices in the magnetic field at the magnetosheath above Venus' ionopause might originate from nonlinear waves at the planetary obstacle boundary [37]. Several theoretical analytical, magnetohydrodynamic (MHD), and kinetic studies on non-magnetized planets such as Venus and Mars also indicate that plasma instabilities which evolve into vortices and possible detached ionospheric clouds can develop [31–33, 38, 39].

The evolution of the KH-instability around the martian ionopause boundary was recently studied by numerical MHD simulations which used realistic atmospheric and ionospheric input parameters at the ionopause boundary layer [40–42]. Figure 3.6 shows the development of the KH-instability by using a normalized mass density at different times during one of these simulations. One can see that after a linear growth time of the instability, a regular-structured plasma vortex can evolve and develop to an nonlinear condition. For the particular case, the ion density in the plasma layer below the ionopause boundary is ∼10 times the density compared to the plasma density above. The results of these studies indicate that a larger density jump stabilizes the boundary layer. From these results one finds that the martian ionopause should be most likely stable with regard to the KH-instability because of the stabilizing effect of the large mass density in the ionosphere [42].

For high solar activity, however, the induced magneto-ionopause of Venus or the magnetic pile-up boundary at Mars might be unstable against the KH-instability. In the martian case, the magnetic pile-up boundary is too far above the ionopause so that the atmospheric loss of planetary ions might not be as efficient as if the ionopause is the unstable boundary. For this reason ion escape from Mars due to plasma instabilities may be a less efficient non-thermal atmospheric escape process than previously thought [42].

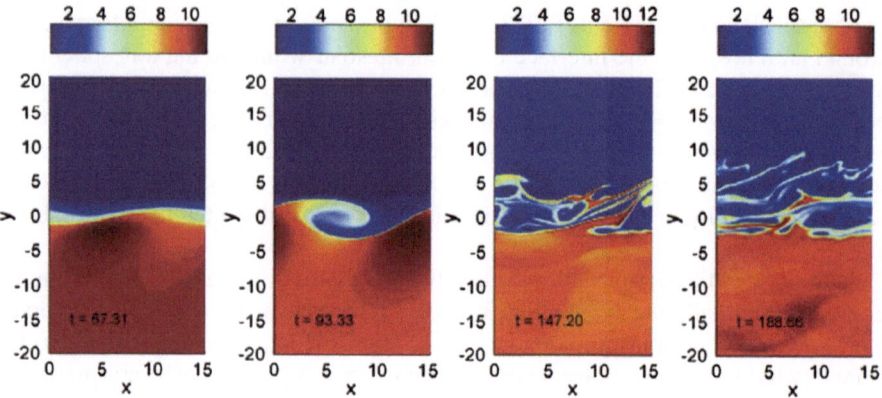

Fig. 3.6 Numerical MHD simulations which show the nonlinear evolution of the-KH instability. The time series of the mass density is shown, from an MHD simulation with periodic boundary conditions in the x-direction where from initially small perturbations a KH-vortex and later a chaotic regime develops. The mass density changes from the upper to the lower plasma layer and exhibits an increase of up to ten times (*blue* low density, *red* high density) (courtesy of U. Möstl)

3.1.5 Momentum Transfer, Cool Ion Outflow, and Polar Wind

In addition to ion pick up and plasma instabilities, two additional non-thermal atmospheric escape processes which are not so well investigated so far should also be briefly discussed. The first process is related to the transfer of momentum from the solar wind to the ionopause boundary layer and the upper ionosphere, which can result in an acceleration of planetary ions and consequently, if the obtained velocities exceed the escape velocity of the planet, in the loss of these ions from the upper atmosphere.

The energy and momentum transfer caused by the incoming plasma flow occurs on the ionized part of the upper atmosphere, where the plasma flow can interact with the cool plasma of the ionosphere and the neutral gas. The plasma interaction around the ionopause boundary leads to energization and as illustrated in Fig. 3.4 to an outward flow of planetary ions throughout the tail [7]. However, a detailed theory on the mechanism which is behind an effective transfer of solar wind momentum to planetary ions below the ionopause is not developed so far.

According to a commonly accepted hypothesis the transfer of solar wind energy and momentum to the ionospheric plasma which is embedded in the flow is expected to continue until a balance in the momentum flux between the solar and planetary tail plasma can be reached.

The solar wind energy and momentum transfer responsible for the energization and escape of ionized atmospheric particles have been roughly described [7, 29, 43–45]. The momentum conservation of energy in the transfer of the solar wind energy and momentum flux F_{sw} and the flux of planetary ions F_{ion} can be written as

$$F_{ion} = \frac{v_{sw} m_{sw}}{v_{ion} m_{ion}} \left(F_{sw} - \frac{v_{loc,\,sw}}{v_{sw}} F_{loc,\,sw} \right) \frac{\xi_{sw}}{\xi_{ion}}, \qquad (3.15)$$

with $F_{loc,\,sw}$ and $v_{loc,\,sw}$ the local decelerated solar or stellar wind plasma flow and velocity inside the interaction region, while v_{sw} and v_{ion} are the solar and stellar wind plasma and planetary ion velocity, respectively. The ratio ξ_{sw}/ξ_{ion} is defined as the relative momentum exchange thickness which can be assumed ≤ 1 [7, 43, 44].

From the observed flow characteristic, data of H^+ or H_2^+ ions in the tail region of Venus from PVO indicate that these ions are accelerated by a polarization electric field [35, 46–48]. Such a polarization electric field force which accelerates cold slow moving ions in the tail are produced because electrons are more mobile than heavier ions and can easily escape from the ionosphere. This polarization electric field force can be written as

$$eE = -\frac{1}{N} \frac{\partial P_e}{\partial z} - \frac{1}{8\pi N} \frac{\partial B^2}{\partial z}, \qquad (3.16)$$

where E is the ambipolar electric field, e the electron charge, the electron pressure P_e, and the magnetic pressure $B^2/8\pi$.

The analysis of the PVO data indicates that the electric field is strong enough to accelerate atmospheric ions throughout the Venus ion tail up to escape energies to an altitude in analogy with the terrestrial ion exosphere. By introducing an interaction cross-sectional area IA one can estimate the net mass loss rate L_{ion} by momentum transfer in units of $g\,s^{-1}$ from a planet as

$$L_{ion} = I A\, m_{ion} F_{ion}. \qquad (3.17)$$

The upper boundary of this interaction cross-sectional area can be assumed as the ionopause location at the terminator, while the lower boundary may be close to the ionospheric peak. One can see from Eq. (3.15) that the planetary ion flux F_{ion} is strongly coupled and connected to the ion velocity v_{ion} and the ion mass m_{ion} of the outward flowing planetary ions. One can also see that a large ion flux corresponds to slower moving ions. Thus, slowly escaping planetary ions, so-called cold ions, may lead to higher atmospheric mass loss rates. Theoretical studies which investigated this ion escape process by assuming larger solar wind and ionospheric parameters as expected through the young Sun or star period indicate that high escape rates can occur.

In case a planet having a strong intrinsic magnetic dipole moment the magnetic field such as that of the Earth, it prevents its atmosphere from the erosion by the solar wind. But due to the partial ionization of the upper atmosphere by the Sun's short-wavelength radiation, electrodynamic forces can accelerate and move ions upward, against gravity, along open field lines within the polar auroral oval [7, 49]. Besides the classical ion acceleration mechanisms by electrical fields, other processes which can move ions upward have also been identified. Another main process besides electrical field acceleration is the transverse heating of ionospheric ions. Due to the locally enhanced plasma pressure, energized ions are pushed upward into the direction of lower magnetic field strength. Related to these plasma disturbances intense small-scale field-aligned currents and electric fields play important roles especially during magnetically disturbed periods [7, 29]. The ion outflow over a planet's magnetic polar areas is called polar wind and is like all atmospheric escape processes strongly coupled to the solar or stellar activity.

3.1.6 Photochemical Escape and Formation of Suprathermal Atom Coronae

The photochemical production of suprathermal or so-called "hot" atoms in the upper atmosphere of terrestrial planets as illustrated in Fig. 3.7 is an efficient process for the origin of excited neutral atoms [50–63]. The most important sources for suprathermal O atoms in the atmospheres of Venus, the Earth, and Mars are

$$O_2^+ : O_2^+ + e^- \rightarrow O(^3P, \, ^1D) + O(^3P, \, ^1D, \, ^1S) + \Delta E, \tag{3.18}$$

for suprathermal C atoms

$$CO^+ : CO^+ + e^- \rightarrow C(^3P, \, ^1D) + O(^3P, \, ^1D, \, ^1S) + \Delta E, \tag{3.19}$$

$$CO : CO + h\nu \rightarrow C(^3P) + C(^3P) + \Delta E, \tag{3.20}$$

while the sources for suprathermal N atoms are

$$N_2^+ : N_2^+ + e^- \rightarrow N(4^S, \, 2^D) + N(^4P, \, ^2D, \, ^2P) + \Delta E, \tag{3.21}$$

$$NO^+ : NO^+ + e^- \rightarrow N(^4S, \, ^2D) + O(^3P, \, ^1D) + \Delta E, \tag{3.22}$$

$$N_2 : N_2 + h\nu \rightarrow N(^4S) + N(^4S, \, 2^D, \, 2^P) + \Delta E, \tag{3.23}$$

and the photochemical reactions for suprathermal H atoms which form non-thermal hydrogen coronae are

$$O^+ + H_2 \rightarrow OH^+ + H^* + \Delta E, \tag{3.24}$$

$$OH^+ + e^- \rightarrow O + H^* + \Delta E, \tag{3.25}$$

$$CO_2^+ + H_2 \rightarrow CO_2H^+ + H^* + \Delta E, \tag{3.26}$$

$$CO_2H^+ + e^- \rightarrow CO_2 + H^* + \Delta E, \tag{3.27}$$

Fig. 3.7 Illustration of dissociative recombination produce suprathermal or so-called hot neutral atoms in the thermosphere. The newly generated hot atoms can move upwards to the exobase level, during their way they will collide with the background gas, change the direction, and lose energy. Those who reach the exobase with velocities $\geq v_{esc}$ will escape from the planet, while those with velocities $< v_{esc}$ will enter ballistic trajectories

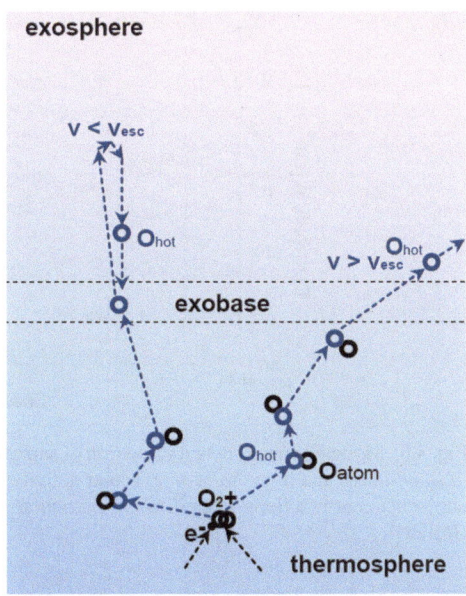

$$H^+ + O \rightarrow H^* + O^+, \tag{3.28}$$

$$H^+ + H \rightarrow H^* + H^+. \tag{3.29}$$

The energy range of these newborn suprathermal atoms depends on the particular photochemical reaction and lies within a range of several 0.1 to <10 eV. The production of suprathermal atoms in the thermosphere and their related energy density function at the exobase can be modeled with Monte Carlo and test-particle models [35, 53, 58, 64–66]. In such models, the velocity distributions of newly created suprathermal atoms is obtained by assuming that the planetary ions and electrons are corresponding to Maxwell distributions according to their altitude depended temperature values. Velocities from these distributions are taken randomly and transformed into the center of mass system, where the velocity direction of the newly born suprathermal atoms is supposed to be isotropically distributed. Because the mass of the electrons is much smaller compared to the heavy suprathermal atoms, the kinetic energy in the center of mass system is dominated by the motion of the electrons and is added to the dissociation energy of a specific branch. Assuming that the energy is equally shared among the two resulting suprathermal atoms, the energy of these species can be found by a back transformation to the initial frame [54, 67]. By taking a large number of randomly chosen velocities for both molecular ions and electrons, the initial energy distributions are obtained. After the production of suprathermal atoms, the particles are subject to binary collisions with the main neutral atmospheric constituents, i.e., O, CO_2, CO, and N_2.

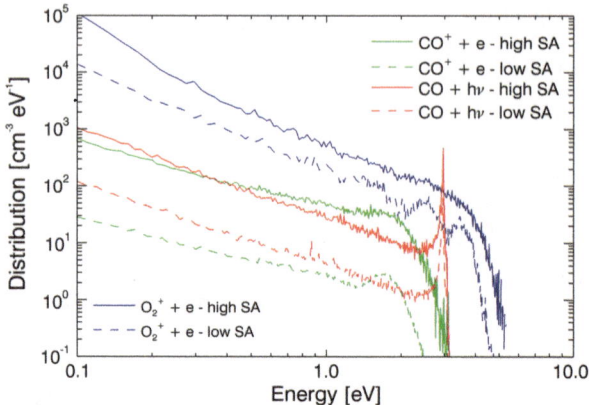

Fig. 3.8 Modeled energy density function of suprathermal O and C atoms at 240 km altitude for low (*dashed-lines*) and high (*solid-lines*) solar activity conditions on present Mars. The corresponding photochemical reactions related to the particular suprathermal atoms are shown (courtesy of H. Gröller)

Figure 3.8 shows the result of modeled energy density functions for suprathermal O and C atoms on present Mars during low and high solar activity by applying a sophisticated Monte Carlo model of [65, 66] which includes

- energy and mass depended total collision cross-sections [59, 60, 68–70] to calculate the collision probability,
- energy- and mass-dependent differential collision cross-sections to determine the scattering angle after a collision [70],
- elastic, inelastic, and quenching collisions, as well as the energy transfer during a collision,
- radiative lifetimes of excited particles,
- production of secondary suprathermal atoms.

The newly born suprathermal atoms are traced up to the exobase or higher altitudes, where their energy density distributions and the corresponding escape fluxes can be calculated. One can see from Fig. 3.8 that these suprathermal atoms are more energetic compared to the atmospheric bulk gas, so that those which have velocities $v \geq v_{esc}$ will escape directly from the planet, while atoms with $v < v_{esc}$ are gravitationally bound to the planet will produce extended suprathermal atom coronae, such as shown in Fig. 3.9. If the ballistic corona particles are not protected by a magnetosphere, they will interact with the solar- or stellar wind plasma and can be eroded from the planet via ion pick up. In case that some of the newly born ions from such a corona [71, 72] are backscattered toward the planet they can act as so-called sputter agents.

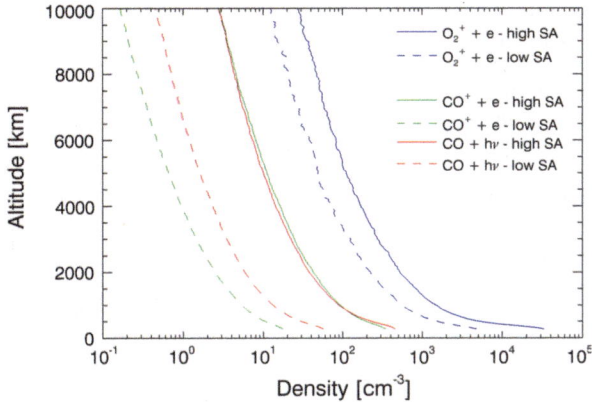

Fig. 3.9 Modeled suprathermal O and C atom corona as a function of particular photochemical production processes at present Mars for low (*dashed-lines*) and high (*solid-lines*) solar activity conditions (courtesy of H. Gröller)

3.1.7 Atmospheric Sputtering

Atmospheric sputtering caused by solar- or stellar-wind protons or backscattered exospheric pick up ions is a collision-based energy transfer process, which can result in ionization or excitation of atmospheric neutral particles (see Fig. 3.10) [73]. The sputtering yield of atmospheric particles can be defined as the number of atoms or molecules which are ejected via collisions with ions or energetic neutral atoms with velocities $v \geq v_{\text{esc}}$ close to the exobase level. This sputter yield includes particles which are ejected directly in a collision with the sputter agent and those ejected due to a cascade of collisions.

The loss rate in units of s^{-1} for atmospheric sputtering L_{sp} from the upper atmosphere of a planet which is not protected by a strong intrinsic magnetic field if the incident particles are ions can be written analytically as [73, 74]

$$L_{\text{sp}} \approx F_{\text{ion}} \left(2\pi r_{\text{exo}}^2\right) \left[\frac{0.5\sigma_{(E_T \geq E_{\text{esc}})}}{\sigma_D \cos\theta} + \frac{3\hat{\alpha} S_n}{\pi^2 (\cos\theta)^{1.6} E_{\text{esc}}\sigma_D}\right], \qquad (3.30)$$

where F_{ion} is the incident ion flux, θ is the pitch angle of the incident particle, E_{esc} is the escape energy of the particle at the exobase, $\sigma_{E_T > E_{\text{esc}}}$ the collision cross-section for a particle which receives an energy transfer, σ_c is the collision cross-section with the sputtered atom or molecule and S_n is the so-called stopping cross section. The constants $3/\pi^2$ and $\hat{\alpha}$ are obtained from the transport equation and the factor α depends on the mass fraction m_B/m_A, where m_B is the target particle mass and m_A is the mass of the incident sputter agent. Atmospheric escape by sputtering is relevant mainly at planetary bodies with lower gravity such as Mars or Titan [4, 74, 75]. However, it is important to point out that strong induced or intrinsic magnetic fields will decrease atmospheric sputtering [76].

Fig. 3.10 Illustration of atmospheric sputtering. Energetic or fast neutral or charged particles which enter the thermosphere will collide with low energetic atmospheric atoms and transfer energy to them. If the transferred kinetic energy obtained by the target atom is efficient enough that its velocity is $\geq v_{esc}$ it will escape

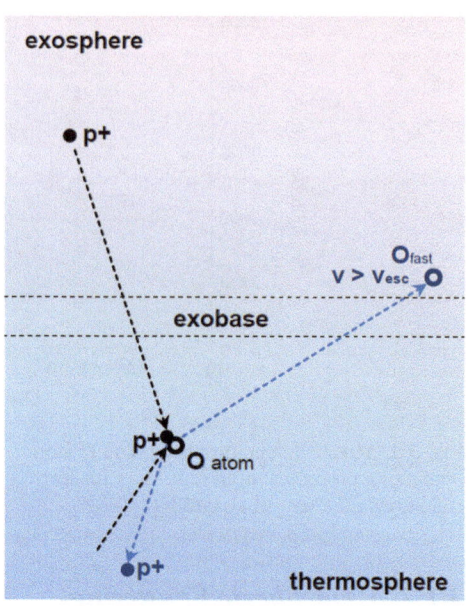

3.1.8 Evolutionary Relevance and Efficiency of Atmospheric Escape Processes

All thermal and non-thermal atmospheric escape processes which are briefly discussed before are strongly coupled on the history of the planet's host star's EUV radiation and on the efficiency of plasma outflow (i.e., winds, CMEs, SEPs, etc.). Figure 3.11 illustrates the major atmospheric escape processes and their efficiency as a function of age of a planet's host star. The influence of a possible intrinsic magnetic moment is also addressed. The first phase and most efficient atmospheric escape period starts after the nebula dissipated and the protoplanet is exposed to the extreme radiation and plasma environment of the young star. During this period, for a planet with the Earth's gravity within the habitable zone of its host star thermal escape of light atmospheric species such as H, D, H_2, HD and He by hydrodynamic blow-off is the major atmospheric escape process. The time period of this extreme escape process depends generally on the evolving EUV flux of a planet's host star, the planet's orbit location, its gravity, and the main atmospheric constituents in the upper atmosphere.

The second stage in atmospheric escape is a transition phase where the upper atmosphere is not in hydrostatic equilibrium, then the thermosphere expands dynamically, but the outward flowing gas does not reach the escape velocity at the exobase level. In this stage the upper atmosphere will experience high Jeans escape and non-thermal atmospheric loss from the expanded thermosphere-exosphere region by ion pick up. For planets with lower gravity in absence of a strong magnetic field photo-

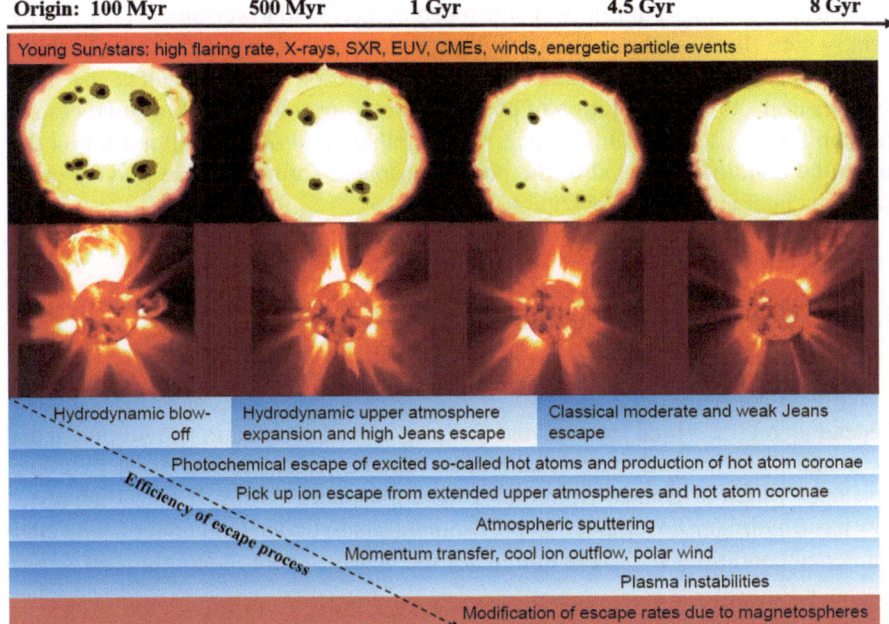

Fig. 3.11 Illustration of various thermal and non-thermal atmospheric escape processes and their efficiency at terrestrial planets. The age of the host star and planet as well as the host star's activity is considered by the *time arrow* from *left* to *right*. After the planets origin, EUV-powered hydrodynamic blow-off occurs most likely and changes with decreasing EUV from hydrodynamically expanded thermospheres into a hydrostatic regime. At the period where this transformation occurs, various non-thermal atmospheric escape processes start to work and contribute to the total atmospheric loss. After the EUV flux of the young and active host star decreases to moderate levels (<5 EUV) and CO_2 or N_2 may get the dominant species in the thermosphere all atmospheric escape processes will contribute to the atmospheric loss but with much lower or even negligible consequences to the atmosphere compared to the early EUV active phase of the young host star

chemically produced suprathermal atoms, sputtering and cool ion outflow can also contribute to the atmospheric escape.

During these early stages of atmospheric escape even an intrinsic magnetic moment of a terrestrial planet may not protect the atmosphere against escape [16]. The upper atmosphere will most likely expand above the corresponding magnetopause where it can be eroded by the solar/stellar-wind plasma flow around the planetary obstacle. This obstacle originates most likely slightly above the exobase level, due to a pressure balance between thermal ions, induced and possible intrinsic magnetic fields against the solar/stellar-wind plasma ram pressure.

The third phase in atmospheric escape occurs when the solar- or stellar EUV flux decreases to more moderate fluxes which are comparable to that of today's solar cycle. Depending on the magnetization and gravity of a planet all escape processes will work but in a more moderate or even negligible efficiency compared to the period of the young active Sun/star.

If one applies these in Fig. 3.11 illustrated scenarios to lower mass K- or M-type stars, where there is observational evidence that their X-ray, EUV energy fluxes remain longer active [77] the related atmospheric escape processes will have a much stronger impact in the evolution of planetary water inventories and in some case on the stability of its whole atmosphere. On the other hand one should not forget that the initial conditions related to the protoatmosphere formation also plays a major role in its later atmosphere evolution.

3.2 EUV-Powered Escape of Hydrogen-Rich Protoatmospheres

Depending on the orbital location and the host stars EUV radiation environment and the planet's gravity, early terrestrial planets may be surrounded by dense hydrogen envelopes which originated from their protoatmospheres. Thus, if one is interested how these hydrogen envelopes escape, its important to understand under which environmental conditions hydrogen dominates the thermosphere in atomic form or as a molecule.

3.2.1 The H_2 Breakdown Region

Photodissociation of H_2 molecules in Jupiter-type hydrogen atmospheres at various closer orbit locations of a Sun-type host star have been studied carefully during the past years [78, 79]. These studies found that hydrogen molecules dissociate when the planet orbit at locations ≤ 0.2 AU. At orbital locations > 0.2 AU, H_2 remains the dominant species in a hydrogen-rich thermosphere. From these studies one can assume that for EUV fluxes > 25 times than today's solar value the majority of the H_2 molecules will be photodissociated and H atoms dominate in the upper atmosphere. The results of these studies are also connected to the photodissociation rate of H_2 and the resulting H^+ density.

If a protoplanet is initially surrounded by a dense molecular hydrogen envelope the high EUV flux of the young Sun/star will dissociate the H_2 molecules via the following reactions [78, 79]

$$H_2 + h\nu \rightarrow H^+ + H + e^-, \tag{3.31}$$

$$H^+ H_2(\nu \geq 4) \rightarrow H_2^+ + H, \tag{3.32}$$

where $H_2(\nu \geq 4)$ corresponds to an vibrationally excited hydrogen molecule. Additional photochemical reactions such as $2H_2^+ + e^-$, $H_2^+ + e^-$ or $H_2 + M$ followed by fast conversation of H_2^+ into H_3^+ and subsequently dissociative recombination forms also preferentially H atoms. For EUV fluxes ≤ 25 times compared to the present

Table 3.1 Exobase distance and temperature of a hydrodynamic expanding H_2 and H-rich thermospheres of an Earth-like planet, exposed to EUV flux values which are 5–100 times higher compared to that of the present Sun

EUV/EUV$_{Sun}$	5	10	30	50	70	100
H_2: r_{exo} (r/r_{Earth})	7	7.7	9.6	–	–	–
T_{exo} (K)	525	625	975	1,225	1,600	1,875
H: r_{exo} (r/r_{Earth})	9.5	10.5	14	16	17.5	19
T_{exo} (K)	360	485	960	1,390	1,745	2,310

[a]For EUV fluxes >25–30 times EUV$_{Sun}$ most H_2 molecules dissociate

solar value, the H^+ density is too small that the above-mentioned neutral chemical reaction are insignificant.

The second possibility to break H_2 molecules into two atomic hydrogen atoms is thermal dissociation [79]

$$H_2(T \geq 2,000\,\text{K}) \rightarrow H + H. \tag{3.33}$$

In case an upper atmosphere is dominated by molecular hydrogen and the temperature of the gas reaches values which are $\geq 2,000\,\text{K}$, the H_2 molecules break down thermally into hydrogen atoms. Table 3.1 shows the exobase distance and temperature corresponding to the input values of Fig. 3.3 for a H_2 (<30 EUV) and H-dominated upper atmosphere for EUV flux values which are 5–100 larger compared to today's solar value at an Earth-mass and size planet. One can see that due to the low gravity of an Earth-like planet the exobase level moves to further distance but cools adiabatically due to the dynamical expansion [15] so that thermosphere temperatures $\geq 2,000\,\text{K}$ can only be reached in the absence of IR-cooling molecules and for EUV fluxes which are ≥ 80 times that of the present Sun. For more massive super-Earths due to their higher gravity the thermal dissociation breakdown temperatures are easier obtained. However, at the time when hydrogen-rich protoatmospheres originated, the EUV flux of the young Sun or young stars inside the habitable zone is ~ 80–100 times higher compared to the present solar value, which is enough to dissociate the molecular hydrogen gas in the upper atmosphere of young planets.

In case an early terrestrial planet is surrounded by a dense steam atmosphere, H atoms should also be the dominant species in the upper atmosphere. Below $\sim 50\,\text{km}$ the solar/stellar UV- and EUV radiation which dissociates H_2O molecules

$$H_2O + h\nu \rightarrow OH + H, \tag{3.34}$$

$$OH + H_2 \rightarrow H_2O + H, \tag{3.35}$$

practically does not penetrate. If one assumes that the water vapor mixing ratio at $\sim 50\,\text{km}$ has more or less a constant value for different solar EUV and UV radiation fluxes, then for the UV- and EUV fluxes higher than for today's Sun, less water vapor can be expected to be present at altitudes above ~ 60–70 km due to stronger

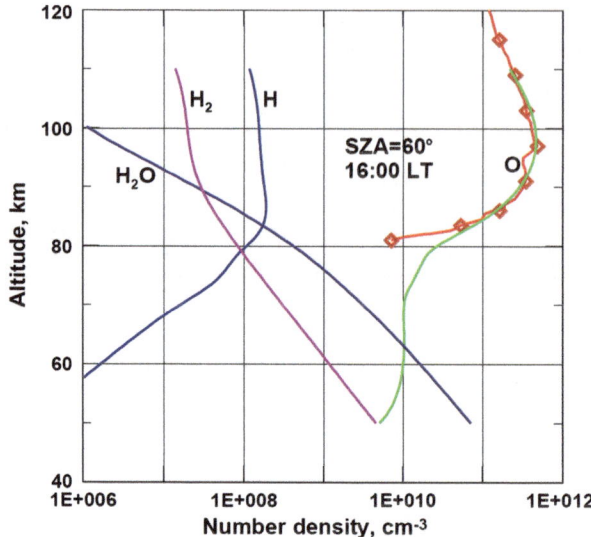

Fig. 3.12 Density profiles of water molecules and its EUV dissociated products such as H_2 and H in the Earth's atmosphere for present solar EUV flux conditions (courtesy of Yu. N. Kulikov)

photodissociation and ionization. This decrease in H_2O molecules will be stronger altitude dependent for stronger EUV fluxes. One can see from Fig. 3.12 that even under the present time EUV radiation exposure the H_2 number density is one order of magnitude lower than the density of atomic H above $\sim$100 km. A faster decrease with the increasing altitude of the H_2 number density can be expected above $\sim$50 km for stronger EUV fluxes, while the atomic H number density is expected to remain nearly the same in the mesosphere-lower thermosphere region if one assumes a constant H_2O mixing ratio at 50 km.

From these brief discussion on H_2 molecule breakdown and H_2O photodissociation one can conclude that due to the high EUV flux of the young Sun/stars and other extreme environmental conditions such as frequent impacts, hydrogen-rich upper atmospheres of young terrestrial planets will most likely be composed in atomic form.

3.2.2 Escape of H-Rich Gas Envelopes

In previous studies related to the escape of nebula-based protoatmospheres, several researchers concluded that nebula-based hydrogen envelopes may be lost after a few tens of Myr [80–83]. However, one should note that in these pioneering studies, which were carried out during the eighties, no observational data of very young solar proxies as shown in Table 2.3 were available [84]. Therefore, applied EUV flux enhancement factors to protoatmosphere escape studies during time periods between $\sim$3–100 Myr after the Sun's origin, overestimated the atmospheric mass loss by orders of magnitudes [80, 85]. For time periods after the first 100 Myr, these studies yielded more or less accurate results. As discussed in Chap. 2, since these early studies, astrophysical observations of solar proxies with younger ages revealed that the EUV flux of these young stars are saturated at an enhancement value of $\sim$100 during the first $\sim$100 Myr after the Sun/star arrives at the ZAMS [86] and decreases then later according to power laws such as that given in Eq. (2.4) [87, 88].

Some studies explained the escape of nebula-based hydrogen envelopes of an equivalent hydrogen amount of several hundred Earth oceans (EOs) during $\sim$20 Myr by assuming that in addition to the high EUV flux, the far-UV radiation was also up to 100–1,000 times higher in the wavelength region between 145 and 185 nm, compared to that of the present Sun [83]. However, as one can see from Table 2.3 astrophysical observations of solar proxies with very young age reveal that the far-UV flux of the young Sun was most likely <20 times during the first 100 Myr after the Sun's origin. However, higher UV values and also X-ray heating could have played a role as long as the young Sun went through its T-Tauri phase. The average T-Tauri phase of most stars last between $\sim$1–10 Myr with its most active period $\leq$3 Myr [89–91]. This period correlates more or less with the systems nebula lifetime [88, 92]. During this very early active phase of the young star the decreasing nebula density will protect the accumulated gas around the protoplanets as long as the gas density is lower compared to the gas envelope. Because this accumulated gas envelopes are not completely separated from the evaporating nebula the escape processes from the upper atmosphere of an nebula-embedded protoplanet will not be as efficient compared to the period when the optical depth of the nebula is negligible and the gas can escape freely from the exosphere. In recent studies, the importance of X-ray heating was pointed out [93, 94] because it was found that hydrogen-rich exoplanets at orbital locations $d < 0.1$ AU are evaporating in two distinct regimes: X-ray-driven, in which the X-ray heated flow contains a sonic point, and EUV-driven, in which the X-ray region is entirely sub-sonic [94]. For X-ray luminosities $L_{X\text{-ray}}$ which are $\geq 10^{29}$ erg s^{-1} related photo-electric heating could heat a hydrogen-rich thermosphere to sufficiently high temperatures that hydrodynamic blow-off occurs [93]. The mass-loss rates scale as $L_{X\text{-ray}}/d^2$ for X-ray driven evaporation and may play an important role for planets with lower gravity such as terrestrial planets, super-Earths, and hot Neptunes in orbital locations $d < 0.1$ AU [94]. For L_{X-ray} values which are $<10^{29}$ erg s^{-1}, EUV heating dominates also the thermal escape of the upper atmosphere.

Furthermore, all pioneering studies applied the so-called energy limited[1] approach where these authors assumed that 100 % of the absorbed energy will be transferred to thermal energy overestimate the escape rates most likely by more than 50 %. As discussed before, due to the high EUV flux of the young Sun/star, H$_2$ molecules are dissociated to hydrogen atoms and thus a part of the thermal energy of the

---

[1] Energy limited means that the ratio of the net heating rate to the rate of stellar energy absorption $\eta = 100\%$.

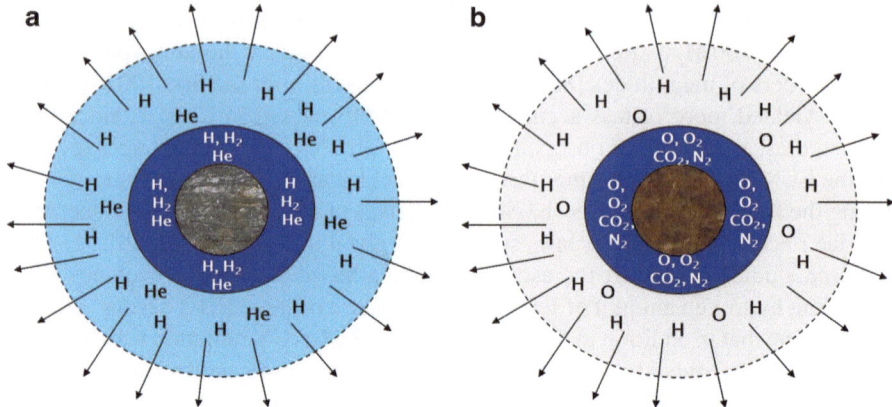

Fig. 3.13 Illustrations of hydrodynamic blow-off of atomic hydrogen atoms from a terrestrial planet which could not lose its solar nebula-based hydrogen/He envelope (**a**) and a planet with a steam atmosphere (**b**). Due to high EUV radiation of the young Sun/star H_2 and H_2O molecules will dissociate and hydrogen atoms will populate the upper atmospheres, where they can drag heavier species such as He or oxygen atoms within the hydrogen wind

Table 3.2 EUV-related heating processes and heating efficiencies η within the hydrogen-rich thermosphere of Jupiter [95, 96]

Heating process	η (%)
Neutral heating	9.26
Electron heating	7.92
Vibrational energy	6.23
Σ	23.41
Chemical heating (H_2^+)	29.74
Total Σ	53.15

gas that could drive atmospheric escape is spent instead on dissociation. Strong photoionization of atomic hydrogen becomes efficient. Furthermore, ion-molecular exothermic reactions, which should produce chemical heating, will in fact not deposit their energy as heat due to a lower frequency of collisions between the reacting species [78]. Additionally, Lyman-α cooling [97] may also influence the heating efficiency so that in the upper layers of the thermosphere η should be in the order of $\sim$10–40 % [13, 14, 78, 95, 97, 98]. A reason for these uncertainties are related to the fact that the heating efficiency changes its value as a function of altitude. It can reach $\sim$60 % in a hydrogen-rich lower thermosphere where most of the energy is deposited and about 10 % in the less dense upper part of a planet's thermosphere. For that reason it is possible that the average heating efficiency in a hydrogen-rich thermosphere which is exposed to high EUV fluxes is closer to $\sim$30–40 % [25, 78].

Thus, if an upper atmosphere is under hydrodynamic blow-off conditions one can apply for the estimation of the atmospheric blow-off loss rate the energy-limited

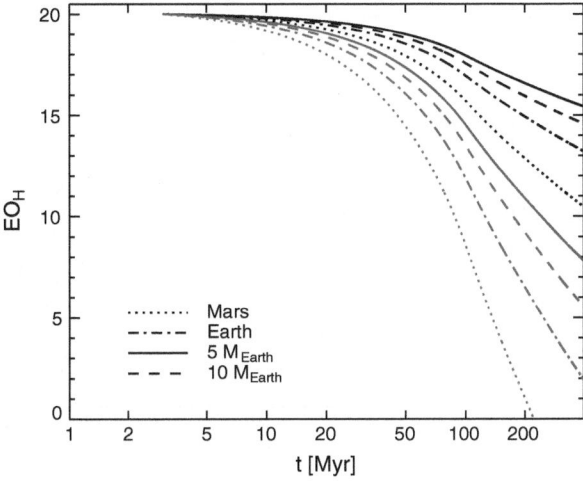

Fig. 3.14 Loss of H atoms in EO equivalents from Mars-, Earth- and super-Earth-type planets at 1 AU, which are exposed during the 200 Myr EUV active phase of a Sun-like star. The *black lines* correspond to heating efficiencies of 15 % and the *grey lines* to heating efficiencies of 40 % (courtesy of P. Odert)

escape formula [99] but have to introduce a heating efficiency η which modifies the energy-limited equation to the following blow-off formula [84, 95, 98]

$$\frac{dM}{dt} = \frac{3\eta S_{EUV} F_{EUV}}{4G\rho_{pl}}, \qquad (3.36)$$

where F_{EUV} is the solar or stellar EUV flux averaged over the planetary sphere, S_{EUV} is an EUV flux enhancement factor according to the power laws discussed in Sect. 2.2 (e.g., Eq. 2.4), G is the gravitational constant, ρ_{pl} the mean planetary density and F_{EUV} the present solar EUV flux at 1 AU in the wavelength range ≈ 2–120 nm [84]. Figure 3.13 illustrates the hydrodynamic blow-off of a nebula-based H-dominated protoatmosphere and a steam atmosphere where the H_2O molecules have been dissociated by an extreme radiation environment of a planet's host star.

Figure 3.14 shows the estimated escape of nebula-based H-rich gas envelopes from a Mars-, Earth-, and two super-Earths with sizes of 1.5 and $2R_{Earth}$ and masses of $5M_{Earth}$ and $10M_{Earth}$ at 1 AU, after a young G-type star arrived at the ZAMS during its most active phase by assuming a conservative η of 15 % and a higher value of 40 %. One can see that planets with low gravity such as Mars can lose hydrogen envelopes more easier compared to larger and more massive ones. However, as one can see even for a higher heating efficiency terrestrial planets and especially more massive super-Earths may have a problem in losing dense hydrogen envelopes via thermal escape. During very early periods the whole atmosphere may be heated by impacts to temperatures which are $\geq 1,500$ K so that the escape may reach energy-limited conditions for a few tens of Myr. The resulted maximum possible atmospheric loss

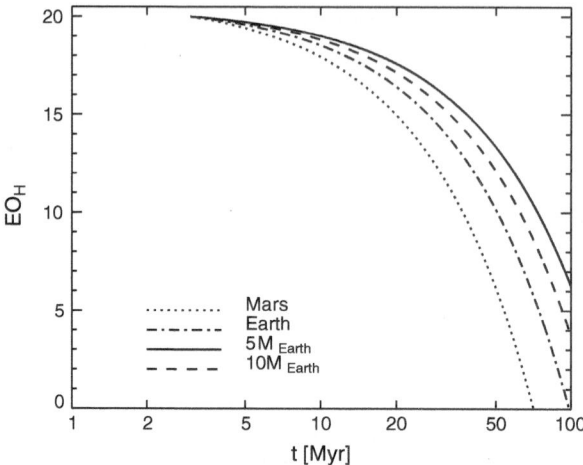

Fig. 3.15 Hydrogen atom loss of similar planets which are exposed to an EUV flux which is 100 times higher compared to that of the present Sun and which experience energy-limited thermal escape at an orbit location of 1 AU during the first 100 Myr after their origin (courtesy of P. Odert)

($\eta \sim 100\%$) of hydrogen for the same planets and an EUV flux which is 100 times that of today's Sun during the first 100 Myr after the origin of the planets is shown in Fig. 3.15. One can see that the early Earth may have lost during that period an amount of hydrogen equivalent of $\sim$20 EOs thermally. However, the more massive super-Earth's may have a problem to lose such hydrogen envelopes thermally. Depending on the formation scenario, nebula dissipation time and available planets in a system, theoretical simulations which have been discussed in Chap. 1 indicate that terrestrial planets could capture much more nebula gas.

Figure 3.16 shows the hydrogen escape of three super-Earths, each of them with a mass of $5 M_{Earth}$ and a size of $1.5 R_{Earth}$ within the orbit of the habitable zone of a Sun-like G-star, which is exposed by a 100 times higher EUV flux since 3 Myr after the origin of their host star. The heating efficiencies are 15 % for the black lines and 40 % for the grey lines. It is assumed that each of these planets captured and accumulated a different mass of hydrogen in units of EO equivalent hydrogen amounts. As one can see these massive bodies would not lose the hydrogen envelope via thermal escape. Due to their high mass only a fraction of the captured hydrogen can escape to space so that these planets finally remain small sub-Neptune-type bodies. If these planets would orbit closer around their host stars ($<$0.1) AU, then they could lose more gas or even the whole hydrogen envelope [98].

From these brief discussions one can expect that the initial phase of the star forming scenario, the time period of the particular extreme active T-Tauri phase, the nebula dissipation time period, as well as the particular planet formation scenario sets very important initial conditions that a planet may further evolve to an Earth-like class I habitat, or ends as a class II or V habitat or remains a non-habitable planet which is surrounded by an accumulated dense hydrogen-rich nebula gas envelope. In

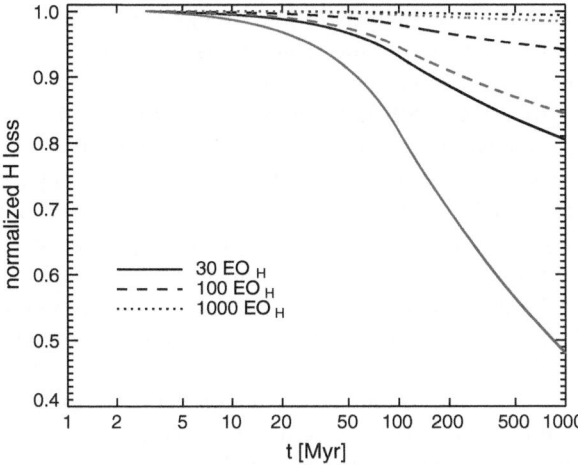

Fig. 3.16 Normalized escape of atomic hydrogen envelopes in units of EO equivalents from a super-Earth within the habitable zone of a Sun-type G-star with a radius of $1.5 R_{Earth}$ and a mass of $5 M_{Earth}$ during the EUV-active period of 1 Gyr after the stars origin. The *black lines* correspond to a heating efficiency of 15 % and the *grey lines* to a heating efficiency of 40 % (courtesy of P. Odert)

the case of the Solar System planets such as Venus or the Earth no hydrogen envelope remained, which indicates that their captured hydrogen from the nebula should have escaped during a few tens or hundreds of Myr after Venus' and the Earth's origin.

3.3 Atmosphere Evolution of Venus, Earth, and Mars

The evolution of terrestrial planetary atmospheres such as Venus, Earth, and Mars after the protoplanets lost their captured nebula-based hydrogen envelopes are strongly linked to the active young Sun after its arrival at the ZAMS, the volatile amount of the catastrophically outgassed steam atmosphere during the solidification of the magma oceans, the thermospheric composition, the amount of IR-cooling molecules, and the mass and size of the particular planet.

3.3.1 Escape of Early Steam Atmospheres

Depending on the formation process and the involved planetesimals and planetary embryos as well as the systems impact history, it is expected that Venus and the Earth may have been surrounded after the magma ocean solidified by a dense water dominated H_2O/CO_2 steam atmosphere with surface pressures ≤ 500 bar [100, 101].

From tungsten W isotope analysis of lunar rocks the accretion age of the Earth can be estimated of $\leq 60\,\mathrm{Myr}$ after the origin of the Sun [102, 103]. Moreover, from planet formation models one can expected that $\sim 80\%$ of the present Earth's mass was accreted at ~ 20–$30\,\mathrm{Myr}$ [80] and $\sim 90\%$ at ~ 30–$40\,\mathrm{Myr}$ after Sun's origin [80]. Although, it is not known when accretion ended on Venus or Mars, one may argue that it may be unlikely that Venus which is in size and mass a comparable planet with the Earth, or Mars which can be even considered as a planetary embryo may have finished their accretion much later. Based on that arguments one can assume that all three terrestrial planets in the Solar System finished their accretion at a time when the solar nebular dissipated and the activity of the young Sun was in the saturation phase with EUV fluxes as shown in Table 2.3 of the order of ~ 100 times from that of the present Sun [84].

As discussed before, due to the high radiation exposure of the young Sun/star as well as the high initial temperature of the steam atmosphere the H_2O molecules will be dissociated. Moreover, during these early times the young planets will also bombarded frequently by meteorites, which keep the atmosphere hot and in steam form by thermal blanketing over tens of Myr [80, 104–108].

It is very likely that atomic oxygen and also carbon atoms can be dragged to space by the outward flowing dynamically escaping hydrogen flux [14, 84, 109]. By studying the fractionation [110] of heavy atoms with the blow-off formula given in Eq. (3.36) one can estimate the loss of hydrogen-dragged heavy atoms such as O and C from early Venus, Earth, and Mars during the EUV saturation period of the young Sun [84, 111]. The flux of the dragged heavy atoms $F_{O,\,C,\,N}$ from the planetary dayside can then be written as

$$F_{O,\,C,\,N} = \frac{3\eta\, S_{EUV}\, F_{EUV}\, X_{O,\,C,\,N}}{8\pi\, r_{sv}\, G\rho_{pl} m_H X_H} \left[\frac{\left(m_H + \dfrac{kT \frac{3\eta S_{EUV} F_{EUV}}{8\pi r_{sv} G\rho_{pl} m_H}}{bg X_H} \right) - m_{O,\,C,\,N}}{\left(m_H + \dfrac{kT \frac{3\eta S_{EUV} F_{EUV}}{8\pi r_{sv} G\rho_{pl} m_H}}{bg X_H} \right) - m_H} \right], \qquad (3.37)$$

with r_{sv} the altitude location in the hydrodynamically expanding thermosphere where the hydrogen atoms reach the sonic velocity, $X_{O,\,C,\,N}$ and X_H are the mole mixing ratios of dragged heavy atoms and atomic hydrogen, $m_{O,\,C,\,N}$ and m_H are masses of the involved particles, b is a molecular diffusion parameter for dragged heavy atoms such as O, C, or N atoms in a hydrogen atmosphere, and T is the average temperature in the lower thermosphere [84, 109–111].

Figure 3.17 shows the escape scenario of H and O atoms for a steam atmosphere with a H_2O amount corresponding to $2\,\mathrm{EO}$ or $\sim 500\,\mathrm{bar}$ for Venus at 0.7 AU by using a lower heating efficiency of 15 % and a higher one of 40 %. Because of the higher and increasing bolometric luminosity of the young Sun at Venus' orbit thermal blanketing of the steam atmosphere most likely inhibited the formation of a liquid water ocean due to its higher surface temperature [106]. One can see that in such an environment an initial water inventory of up to $\sim 2\,\mathrm{EOs}$ could have been lost during

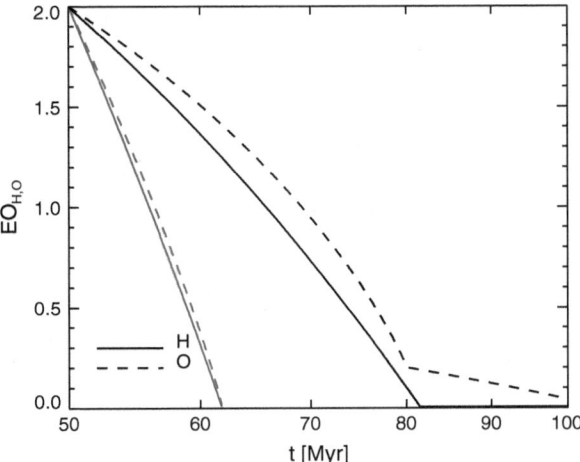

Fig. 3.17 Hydrodynamic escape of atomic H and dragged O from early Venus during the EUV saturation phase of the young Sun, where the early planet's upper atmosphere was exposed with an EUV flux which was ~100 times higher than today's Sun. An upper limit of about 2 EOs (~500 bar) could have escaped during a few tens of Myr during this active period of the young Sun. The *black lines* correspond to a heating efficiency of 15 %, while the *grey lines* are related to a higher heating efficiency of 40 % (see also [84])

the EUV saturation phase of the young Sun. One can also see that for a higher heating efficiency of 40 % the hydrogen drag can remove most of the oxygen after ~12 Myr.

When the amount of available hydrogen atoms decreased, CO_2 was the dominant atmospheric constituent and oxygen in the equivalent amount of several bar to tens of bar could remain in the upper atmosphere. Table 3.3 summarizes the results of O^+ pick up ion loss in units of bar for the minimum, average, and maximum solar wind density values as expected for the young Sun at Venus orbit location [9, 112]. The planetary obstacle in these model runs was assumed near the exobase level of an upper atmosphere which was heated, but remained hydrostatic, by the corresponding EUV flux shown in Table 2.3. That the upper atmosphere of a CO_2-rich early Venus was most likely under hydrostatic conditions is also in agreement with a recent study, where a one-dimensional multi-component hydrodynamic thermosphere/ionosphere model which investigated the thermal and chemical responses of CO_2-dominated upper atmospheres of high EUV exposed super-Earths with 6–10 Earth-masses was applied [12]. The results of this study indicate that the CO_2 atmospheres of the test planets remained in the hydrostatic regime for EUV flux values <200 times that of the present Sun. In case the thermosphere would change to the hydrodynamic regime the exobase would be cooler but expanding to higher a altitudes, which results in a larger solar wind interaction cross-section and higher non-thermal escape rates. Thus, from these studies one can expect that accumulated oxygen in the orders of a few bar to several tens of bar could have been lost from early Venus by non-thermal escape processes such as ion pick up [9]. Figure 3.18 shows the similar escape scenario for H

Table 3.3 Pick up loss of atomic oxygen ions (in units of bar) for the period from 3.6 to 4.5 Gyr before present (b.p.) [9]

t (Gyr) b.p.	3.6	3.8	4	4.2	4.4	4.5
EUV	7	15	30	50	70	100
z_{exo} (km)	~280	~300	~370	~520	~820	~2200
Minimum (bar)	0.015	0.027	0.06	0.15	0.74	4.3
Average (bar)	0.035	0.065	0.16	0.42	2.5	17
Maximum (bar)	0.065	0.12	0.28	1	10	78

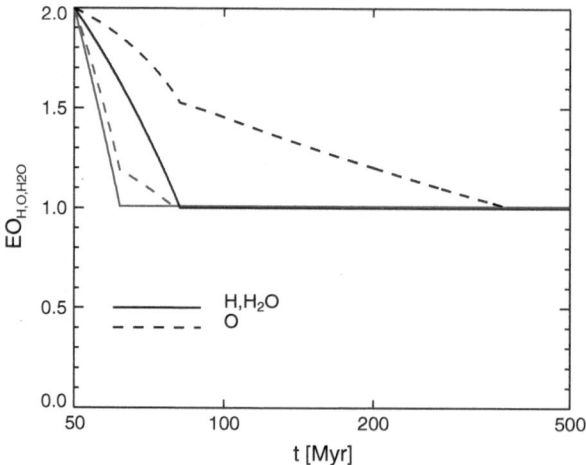

Fig. 3.18 Escape scenario of a catastrophically outgassed steam atmosphere with a water content of up to 2 EOs from the early Earth during the EUV-saturated phase of the young Sun. The *solid-line* shows the escape of atomic hydrogen from the initially outgassed H_2O atmosphere until the equivalent amount of 1 EO is reached. The *dashed-line* corresponds to the dragged O atoms which accumulate to a residual fraction of a corresponding amount of ~0.5 EO when the H blow-off ends due to the formation of the Earth's ocean. The *black lines* correspond to a heating efficiency of 15 %, while the *gray lines* represent the blow-off loss of the steam atmosphere for a heating efficiency of 40 %

and O atoms of a catastrophically outgassed 500 bar (~2 EOs) steam atmosphere for early Earth at an orbit location of 1 AU. Because at that further distance compared to Venus' orbit, the solar luminosity was lower and one can expect that after the loss of about 1 EO (~250 bar) the Earth's atmosphere environment near the surface reached the critical temperature of ~650 K. After reaching this temperature the remaining H_2O-vapor of ~1 EO could condense and collapsed into the liquid water ocean [84]. Additional amount of water could have been delivered also continuously via impacts, but the bulk of the early Earth's initial water inventory is most likely a by-product of a condensed fraction of the catastrophically outgassed steam atmosphere. Similar as on Venus, if a fraction of several bar or up to a few bar of oxygen accumulated in the upper atmosphere it could have been lost during the following few hundred Myr similar as described before by non-thermal escape processes [16].

One can also see that slightly higher heating efficiencies in the order of $\sim$30–40 % result in a stronger hydrogen blow-off flux which drag nearly all remaining oxygen atoms. One should also note that in the case of the early Earth due to the Moon forming impact a fraction of $\leq$30 % of atmosphere could have also been lost to space [113]. Figure 3.19 illustrates the atmosphere evolution scenarios for Venus and Earth after the formation of a catastrophically outgassed water vapor-rich protoatmosphere [100, 114]. On Earth a complex interplay between the young Sun's EUV and plasma environment, atmospheric escape processes, large impacts, and weathering of CO_2 into carbonates during the first 500 Myr resulted in the formation of the Earth's $\sim$0.8 bar nitrogen atmosphere $\sim$3.5–4 Gyr ago [16, 115, 116].

The most actual hypotheses on the formation, outgassing, and evolution of the martian atmosphere since the early Noachian up to the present time was recently investigated [111]. During its formation Mars may have accumulated water and CO_2 equivalent to a range of $\sim$0.06–0.27 times that of an Earth ocean, corresponding to $\sim$20–100 bar [100, 117, 118]. These values are in agreement with recent studies on a catastrophically outgassed steam atmosphere during the magma ocean solidification process on early Mars if the planet originated with initial volatile contents of $\sim$0.05 wt.% H_2O and $\sim$0.01 wt.% CO_2 [100]. The black lines correspond to heating efficiency of 15 %, while the gray lines represent the hydrogen and oxygen escaping atmosphere related to a heating efficiency of 40 %. By assuming a similar accretion age for Mars as for the Earth one finds that due to the high EUV activity of the young Sun and the planet's low gravity, its outgassed protoatmosphere was most likely not stable and escaped via hydrodynamic blow-off [111]. As one can see from Fig. 3.20 depending on the heating efficiency, heavier atoms in the mass domain such as O and C could also be lost by an efficient hydrogen drag during $\sim$5–10 Myr. Impacts which should have occurred during the early Noachian may have kept Mars' protoatmosphere in vapor form but may not have much contributed to atmospheric growth because the delivered volatiles would have also escaped from the planet.

Figure 3.21 illustrates the atmosphere formation and escape of the martian atmosphere during the planet's history. After the martian steam atmosphere escaped, a complex interplay of impact losses, volatile delivery, and mantle outgassing of CO_2 and H_2O may have built up a secondary CO_2 atmosphere which was most likely <1 bar $\sim$3.8–4.3 Gyr ago. However, this secondary atmosphere has to be lost to space or weathered into the surface since that time until present.

3.3.2 Atmospheric Escape from Venus and Mars During Modern Solar Activity

At time periods when the EUV flux of the young Sun decreased to values of $\leq$7 times that of today's Sun thermal escape of heavier species decreased and several non-thermal atmospheric escape processes, which were less relevant during the first 500 Myr after the origin of the planets began to act. The total loss of hydro-

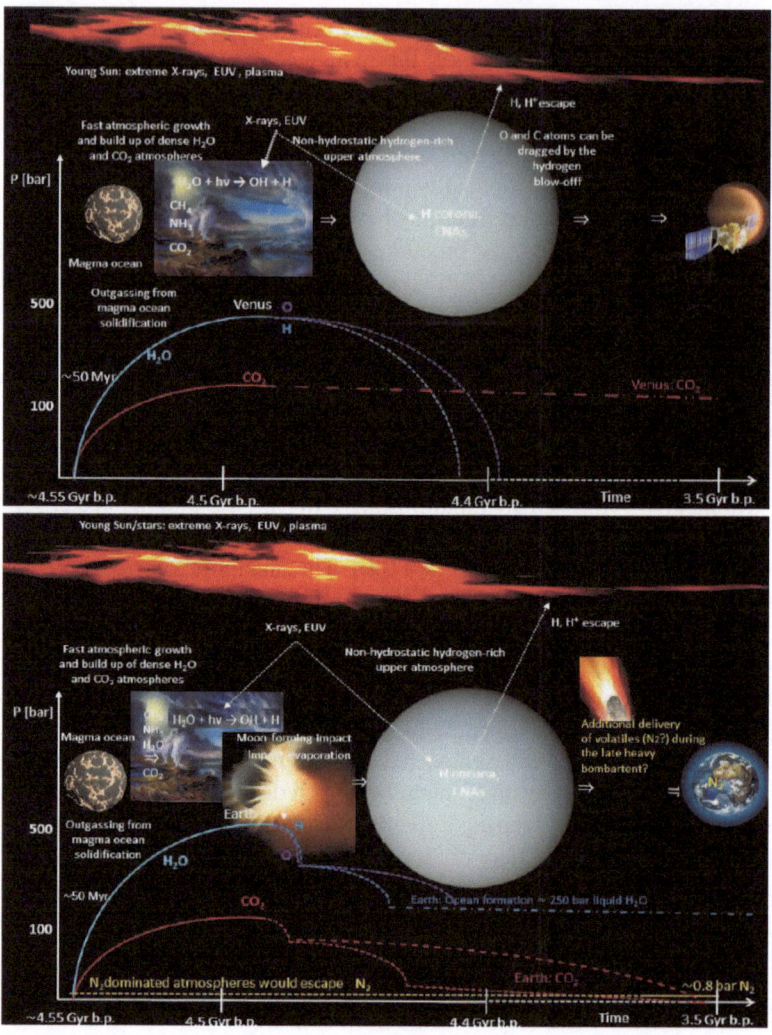

Fig. 3.19 *Upper panel* Illustration of the formation and escape of an early steam atmosphere from Venus. During the evaporation phase of the planets' water inventory, which remained after its outgassing from the solidifying magma ocean most likely mainly in gaseous form, a huge atomic hydrogen corona should have developed from the dissociation of the H_2O molecules. Venus' initial water inventory most likely evaporated during the blow-off phase. *Lower panel* Illustration of a similar scenario for early Earth after the formation of a dense steam atmosphere during the solidification process of the magma ocean [100]. Similar as on early Venus, after the catastrophically outgassed hot H_2O/CO_2-vapor atmosphere the high EUV flux of the young Sun dissociated the H_2O molecules the planet developed a huge hydrogen exosphere with contributions of a solar wind produced ENA cloud. Contrary to Venus, which orbits closer around the Sun, the solar luminosity was weaker so that the atmosphere cooled faster to the level where the remaining water vapor could condense and contributed to the Earth's liquid ocean. CO_2 could then weather out of the atmosphere via early continents and is stored in the lithosphere in the form of carbonates and nitrogen become the dominant species in the Earth's atmosphere. Besides outgassing from volcanoes, volatiles could also be delivered during the late heavy bombardment (see also [84])

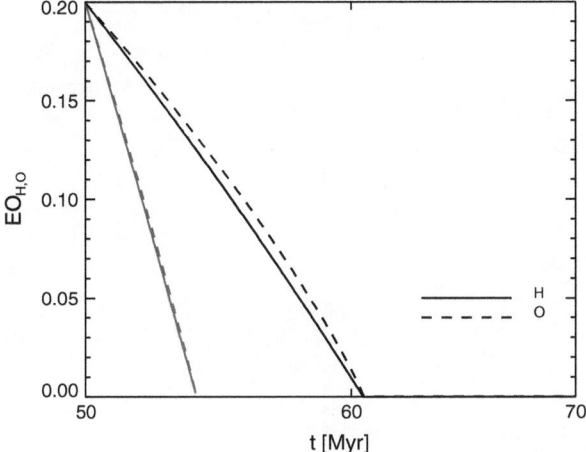

Fig. 3.20 Calculated normalized loss of an outgassed 70 bar (∼0.2 EO) water vapor steam atmosphere from early Mars as a function of time. The *solid-line* corresponds to loss of the hydrogen content, while the *dashed-line* corresponds to dragged oxygen atoms which originate from dissociation of H_2O

gen and oxygen from present Mars and Venus depend on the efficiency of several non-thermal atmospheric escape mechanisms [4, 27, 35, 120]. Compared to a planet with a low gravity such as Mars where photochemical processes result in the production of suprathermal O, C, and N atoms which could overcome the escape energy at the exobase, the escape of particles from the upper atmosphere of Venus is mainly influenced by the escape of hydrogen and oxygen ions caused by the interaction of radiation and the solar wind with the exosphere.

Table 3.4 summarizes modeled and observed average present time hydrogen and oxygen escape rates from both planets.

This comparative study indicates that on Venus, because of its larger mass compared to Mars, atmospheric thermal escape and photochemical processes yield negligible escape rates. Therefore, on Venus ion loss processes caused by the interaction with the solar wind are the most efficient escape processes at present time.

By applying a Monte Carlo model which traces photochemically produced suprathermal atoms [65, 66] from their point of origin inside the thermosphere up to the exobase level for high and low solar activity conditions to present Mars one finds that photochemical processes of heavy atmospheric species are more relevant compared to ion loss processes. Table 3.5 shows the modeled total escape rates for suprathermal O, C, and N atoms during low and high solar activity on Mars. One can also see from the loss rates of escaping heavy main atomic species that the loss rate values depend nonlinearly on the solar activity and related EUV fluxes.

The discovery of water, i.e., deposits below the martian surface polar areas and geological features such as river-type channels indicate that liquid water existed once on the surface in the planet's past. The search and mapping of the present

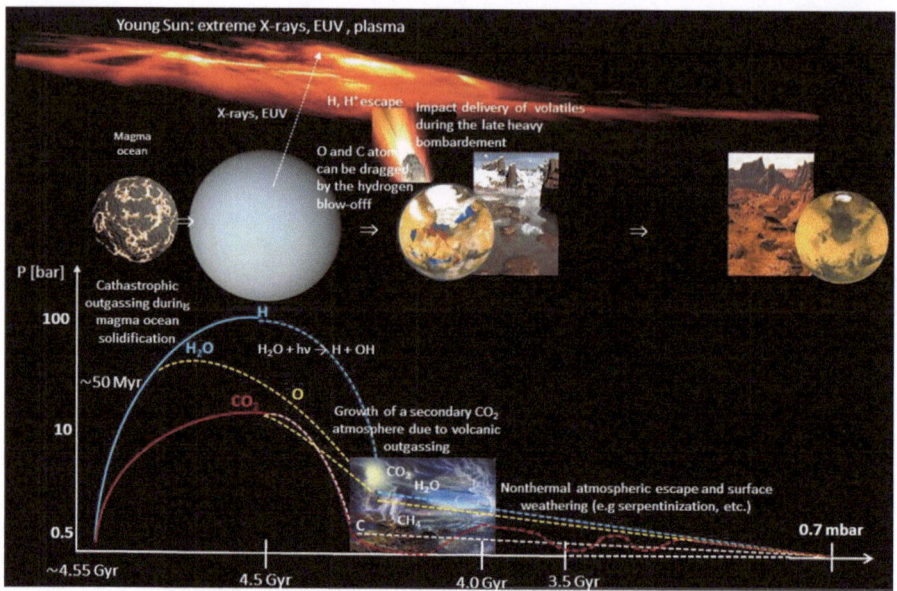

Fig. 3.21 Illustration of Mars' atmosphere evolution after the outgassing of volatiles and fast growth of a dense steam atmosphere during the magma ocean solidification process. Due to the low gravity of Mars one can expect a complex interplay between the young Sun's EUV activity, impacts, and atmospheric escape processes during the first ~500 Myr. After the EUV flux of the young Sun decreased at ~4–4.2 Gyr ago, volcanic outgassing as discussed in Sect. 1.3 and impacts during the late heavy bombardment could have resulted in the growth of a secondary atmosphere CO_2 (<1 bar) [111]. This secondary CO_2 atmosphere should have been lost since the past 3.5–4 Gyr by non-thermal atmospheric escape and surface weathering processes such as serpentinization [119, 111]. On Mars CO_2 can also be released periodically from surface reservoirs during cycles of warmer periods (see also [111])

subsurface water and, i.e., reservoirs is carried out experimentally by ESA's Mars Express and NASA's Mars Reconnaissance Orbiter with their advanced radars for subsurface sounding. By using the observed D/H ratio in the atmospheric water vapor, measured D/H ratios in Martian SNC meteorites and D/H isotope ratios [117] based on asteroid and cometary water delivery to early Mars one estimate a present water, i.e., reservoir which can exchange with the Martian atmosphere equivalent to a global martian ocean layer with a thickness of about 3.3–15 m [120].

By estimating the atmospheric escape rates of water, which correspond to the escape processes related to hydrogen and oxygen atoms shown for present Mars in Tables 3.4 and 3.5 over the past 3.5 Gyr and found that an equivalent global level of water (EGL) to a global martian ocean with a depth of ≤12 m may have escaped to space during this time span. If not all of Mars' initial water inventory, or water delivered during the late heavy bombardment was lost to space, a remaining water content up to ~100 m deep EGL could have been trapped in the martian crust under

Table 3.4 Average escape rates of H, H^+, H_2^+, O and O^+ from Venus' and Mars' present time exosphere

Planet	Venus	Mars
Escape process	Loss rate (s^{-1})	Loss rate (s^{-1})
Jeans: H	2.5×10^{19}	1.5×10^{26}
Photo-chemical reactions: H*	3.8×10^{25}	
Photo-chemical reactions: O*		5×10^{25}
Electric field force: H^+	$\leq 7 \times 10^{25}$	
Ion pick up: H^+	1×10^{25}	1.2×10^{25}
Ion pick up: H_2^+	$< 10^{23}$	1.2×10^{25}
Ion pick up: O^+	$< 10^{25}$	3×10^{24}
Detached plasma clouds: O^+	5×10^{24}–7.5×10^{25}	$\approx 1 \times 10^{24}$
Sputtering: O	5×10^{24}	2.2×10^{23}
Cool plasma outflow: O^+	$< 10^{26}$	

The thermal, photochemical and ion loss rates for Venus are based on model results [35], while the ion and photochemical escape rates for Mars are taken from [27] and [111]. The ion loss rates related to detached plasma clouds correspond to the studies of [39] and [42]. The sputtering escape rates for Venus are taken from [75], the electric force driven H^+ loss rates due to ionospheric holes on Venus nightside are based on estimations by [47], and cool ion outflow is estimated from [121]

Table 3.5 Photochemical (*DR* dissociate recombination, *PD* photodissociation) atmospheric escape rates during low (LSA) and high (HSA) solar activity conditions at present Mars

Elements	O	C	N
DR: L_{HSA}	$\sim 9 \times 10^{25}$	$\sim 1 \times 10^{25}$	$\sim 9 \times 10^{23}$
PD: L_{HSA}		$\sim 1.5 \times 10^{25}$	$\sim 2.5 \times 10^{25}$
Total:	$\sim 9 \times 10^{25}$	$\sim 2.5 \times 10^{25}$	$\sim 2.5 \times 10^{25}$
DR: L_{LSA}	$\sim 3 \times 10^{25}$	$\sim 6.8 \times 10^{23}$	$\sim 1 \times 10^{23}$
PD: L_{LSA}		$\sim 2.8 \times 10^{23}$	$\sim 8.9 \times 10^{22}$
Total:	$\sim 3 \times 10^{25}$	$\sim 9.6 \times 10^{23}$	$\sim 1 \times 10^{23}$

the form of water–ice, hydrated minerals, and liquid water under the cryosphere [119].

What is not understood well so far is the fate of the secondary outgassed CO_2 atmosphere which ended its growth ~ 3.8–4.2 Gyr. From Mars Express Analyzer of Space Plasma and Energetic Atoms-3 (ASPERA-3) and Venus Express ASPERA-4 ion data, one can estimate that only a tiny amount of CO_2^+ molecular ions was lost to space since the end of the Noachian after Mars' magnetic dynamo stopped working equivalent to pressure values of about ~ 0.2–4 mbar [121]. This finding is also in agreement with various theoretical MHD and hybrid modeling results that the planet lost carbon dioxide since the end of the Noachian in the order of ~ 0.008–0.1 bar [4, 111, 122–125].

Table 3.6 shows the expected CO_2 loss in units of bar during the past 4 Gyr, based on the latest spacecraft observations and model studies. One can see that Mars may have lost not more then ≤ 100 mbar of its CO_2 atmosphere to space since

Table 3.6 Expected escape of CO_2 from Mars' atmosphere in units of bar by various non-thermal atmospheric escape processes integrated during the past 4 Gyr [111]

Sources and loss processes	CO_2 (bar)
Ion pick up since ~4 Gyr ago	~0.001–0.1
Detached plasma clouds since ~4 Gyr ago	~0.001
Atmospheric sputtering since ~4 Gyr ago	~0.001–0.1
Dissociative recombination since ~4 Gyr ago	≤0.005–0.05
Photodissociation since ~4 Gyr ago	≤0.005–0.05

the end of the Noachian until present [111]. Thus, if a CO_2 atmosphere of several 100 mbar was indeed present at the end of the Noachian, the majority of it should have been weathered to the crust by sequestration in carbonate rocks and partially recycled to the atmosphere under reduced and/or oxidized form [119]. Depending on the efficiency of these removal processes, this secondary outgassed atmosphere finally evolved to the present time ~7 mbar surface pressure. By comparing the atmospheric escape rates of Venus and Mars one finds that Venus' atmospheric loss of heavy atoms was dominated by non-thermal ion escape processes since the past ~4–4.3 Gyr ago, while on Mars due to its lower gravity photochemically produced suprathermal neutral atoms such as O, C, and N have most likely dominated the atmospheric escape processes.

3.3.3 Ion Escape from the Earth's Magnetospheric Environment

The present average atmospheric mass loss of hydrogen, oxygen, and nitrogen ions from the Earth is $\sim 1.3 \times 10^3$ g s^{-1} [126]. The escape rate depends not only on the solar activity during the solar cycle but also strongly on shorter time scale phenomena such as CMEs, flares, SEPs, etc [127]. There are two main reasons for the enhanced ion outflow during high solar activity: an increased scale height from enhanced X-ray, SXR, and EUV fluxes, and enhanced solar wind plasma forcing. Because the Earth has a magnetosphere, the escape of these atmospheric ions is induced by indirect solar forcing down to the auroral ionosphere guided by the Earth's dipole magnetic field. Because of the strong magnetic dipole field and its resulting magnetosphere, a recycling of outwards flowing planetary ions takes place [7, 128]. As illustrated in Fig. 3.22 the global ion outflow area from the Earth has mainly two source regions, the polar caps and the auroral ovals. For these areas the escape rates are strongly connected to the geomagnetic activity which is described by the planetary K_p-index which quantifies disturbances in the horizontal component of the Earth's magnetic field [129]. Statistical studies suggested that the total O^+ outflow rate is positively proportional to EXP($0.5 K_p$) under the same solar EUV flux value [7]. The escape efficiency due to the outflow rate from the Earth has not been studied very well due

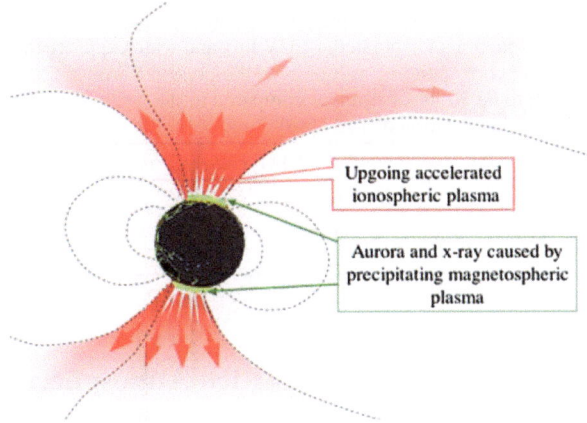

Upgoing accelerated ionospheric plasma

Aurora and x-ray caused by precipitating magnetospheric plasma

Fig. 3.22 Illustration of polar outflow of ions caused by external solar/stellar radiation and plasma forcing on a planet with a strong magnetosphere such as the Earth's (after [7])

to the lack of enough measurements. The total O^+ ion escape rate from the dayside magnetopause and the distant tail is estimated to be about $\sim 5 \times 10^{24} \, s^{-1}$ during low solar activity conditions and of $\sim 4.3 \times 10^{25} \, s^{-1}$ for average conditions [128].

Recently, the total O^+ ion outflow by the analysis of Cluster data and compared the ion escape rates with Mars have been studied [130]. They found from the two intervals when Cluster observed O^+ ions that the total O^+ outflow rate increases by 3–7 times when the K_p-index increased from 0–1 to 3–4. Since the O^+ outflow rate from the cusp/cleft is comparable to that from the polar cap and auroral oval during quiet condition (K_p-index $\approx$ 0–2) [129], the cusp source is expected to become more important when it is observed to increase by ~ 7 times.

Furthermore, corotating interaction regions (CIRs) between the high-speed streams and the ambient solar wind are frequently associated with high solar wind dynamic pressure and with a fluctuating interplanetary magnetic field (IMF), rather than sustained southward IMF [130]. Depending on the magnetospheric configuration and geomagnetic disturbance the escape flux can vary between ~ 37 and 85 %. In the analysis of this study it was concluded that the CIR did not cause strong geomagnetic disturbances [130]. Therefore, it is suggested that the total O^+ escape rate would increase by less than one order of magnitude.

Compared to Mars, the majority of the O^+ ion flow which crosses the martian terminator region will not return back to the planet, because the Larmor radius of heavy ions with energies of tens of eV is larger than the planet's radius [4]. For that reason, under an increased solar wind dynamic pressure increase of ~ 2–3 nPa, the rate of increase in the martian O^+ ion escape flux could be one order higher than on the Earth. At present in the Earth the escape of nitrogen in the form of N^+ and N_2^+ is relatively low compared to the escape rate of O^+ and corresponds to an N^+/O^+ escape ratio which is in the order of ~ 0.1. This result indicates that the loss of volatiles from present the Earth by non-thermal atmospheric escape originates mainly from water.

Table 3.7 Escape of a N_2-rich Earth-like atmosphere from a planet with the size and mass of the Earth in units of bar as a function of solar/stellar EUV flux and wind exposure time Δt^w, for a weak solar wind, and Δt^{st} with corresponding values as at the present Earth and a stronger early solar wind, which is assumed to be 30 times stronger compared to the present solar wind during a time period of 10 Myr at an orbital location of 1 AU [16]

EUV/EUV_{Sun}	$\Delta t^w = 10$ Myr	$\Delta t^{st} = 10$ Myr
20	1 bar	5 bar
10	0.4 bar	3 bar
7	0.2 bar	1 bar

3.4 Stability Problems of Nitrogen-Rich Atmospheres

In a recent studies the response of Earth's upper atmosphere to extreme solar EUV conditions have been investigated [10, 11]. The results of these studies indicate that the lower thermosphere of an Earth-mass planet which is exposed to EUV fluxes which are ≥ 7 times that of the present Sun can be heated to $\sim$7,000–8,000 K. In such a case the upper thermosphere starts to expand but cools adiabatically due to the dynamically outward flowing main atmospheric species (e.g., O, N, etc.). The applied theoretical models of these particular studies are validated against satellite drag observations in the upper atmosphere and the results agree also well with other empirical and theoretical models [8, 131–133].

In their simulations the exobase level moves from present day's location at $\sim$500 km up to about $\sim$2.4, $\sim$4.8 and $\sim$12.7 R_{Earth} above the planetary surface for a solar EUV flux of 7 EUV, 10 EUV, and 20 EUV, respectively. The exobase level which separates the collision-dominated upper atmosphere from the collisionless exosphere is for an Earth-type nitrogen-rich atmosphere which is exposed to about 20 EUV at $\sim$3.7 the Earth-radii above today's average subsolar magnetopause stand-off distance of $\sim$9 R_{Earth} above the surface [16]. Under such extreme conditions the Earth's magnetic moment is too weak to produce a magnetosphere which can protect the exosphere against strong Venus- type non-thermal ion pick up loss processes.

Under solar EUV activity conditions ≥ 10 EUV a nitrogen-rich atmosphere with $\sim$1 bar surface pressure and a composition of the present Earth's atmosphere would have been lost during $\sim$10 Myr (see Table 3.7) [16, 84]. To overcome the stability problem of the early Earth's nitrogen atmosphere inventory it is suggested that a CO_2 amount of at least two orders of magnitude higher than the present time level was needed to confine the thermosphere-exosphere after the onset of the geodynamo within the protecting magnetosphere to avoid it from complete destruction [16].

It is important to note that Earth's atmosphere is not enriched in ^{15}N compared to the solar ^{15}N/^{14}N ratio, it is very likely that its "initial" atmospheric nitrogen inventory has remained unaffected until today [16, 84].

If the early Earth did not lose its outgassed nitrogen inventory during the first 500 Myr after the Sun's arrival at the ZAMS, or it was delivered again during the late heavy bombardment $\sim$3.8 Gyr ago [134], other protection mechanisms should have been active. An alternative protection mechanism for lower abundant heavier atmospheric species would be a dense hydrogen envelope which could have been a remnant of the hydrogen-rich protoatmosphere [84, 135].

If such a dense hydrogen envelope as illustrated in Fig. 3.13 surrounded the early Earth during a few hundred Myr after the planet's origin, the scale height of the EUV heated thermospheric H atoms would be much larger compared to nitrogen molecules which would be located in the lower atmosphere. In such a case nitrogen would not reach the stellar wind interaction region, similar as on the present Earth and as long as hydrogen was the dominant species in the upper atmosphere it would act as a shield against the solar wind erosion of heavier species [136]. However, if the early Earth's initial nitrogen environment was indeed protected by such gas envelopes, these gases should have been lost during a few hundred Myr after the planet's origin.

3.5 Implications for the Search of "Earth-Analogue" Exoplanets

Recent results from ESOs High Accuracy Radial velocity Planetary Search (HARPS) planet finder facility and the discovery of several exoplanets within the so-called super-Earth domain such as Gliese 876d, OGLE-2005-BLG-390Lb, HD 69830b, Gliese 581c, Gliese 581d, CoRot-7b, GJ 1214b, Kepler-10b, Kepler-11b, Kepler-11f and 55 Cnc e [137–145] indicate that planets not much larger but slightly more massive compared to the Earth or Venus are very common in the Universe. From the available statistic one can also expect that many of these planets will orbit inside the habitable zones around faint M-type dwarf stars with about 100 potential target stars in the immediate neighborhood of the Sun [146]. Furthermore, the recent discovery of the first transiting super-Earth candidate, Kepler-22b, with a size of about $2.38 \pm 0.13 R_{Earth}$ within the habitable zone of its solar-type host star at 0.7 AU indicate that it is only a matter of time before the first Earth-size exoplanet inside a habitable zone will be discovered [147]. However, besides super-Earths which are at very close orbital distances (<0.02 AU) such as CoRoT-7b and Kepler-10b with known sizes and masses, the Kepler space observatory discovered several low density planets with a radius-mass relationship which indicates that these planets have rocky cores which are surrounded by a significant amount of H_2 and He envelopes [144].

EUV-powered thermal mass loss calculations over evolutionary timescales for CoRoT-7b and Kepler-10b indicate that hydrogen-rich gas giants within the mass domain between Saturn and Jupiter cannot thermally lose their hydrogen atmospheres that CoRoT-7b or Kepler-10b-type planets result as a rocky residue [95, 98]. However, if one considers also X-ray heating and X-ray-powered mass loss during their earliest evolutionary stage they could have originated as hot Neptunes.

The mean densities of super-Earths such as Kepler-11b and Kepler-11f, of 3.1 and $0.7 \, \mathrm{g \, cm}^{-3}$ [144, 147], and the super-Earth GJ 1214b with a mean density of

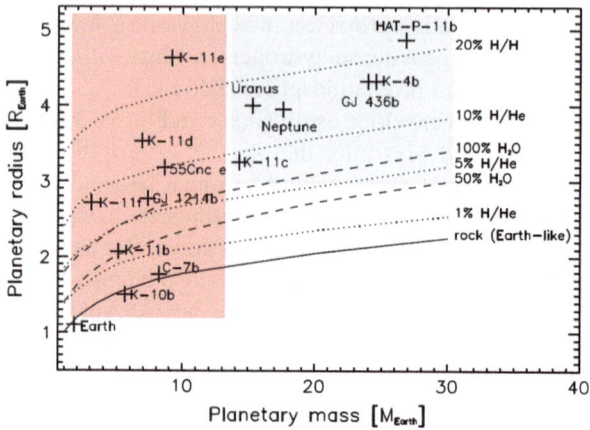

Fig. 3.23 Mass-radius relationship of discovered super-Earths (*red shaded area*) and hot Neptunes (*blue shaded area*) compared to the values of the Earth, Uranus, and Neptune. Kepler planets (K), CoRot-7b (C-7b), GJ 1214b, GJ 435b, HAT-P-11b, and 55 Cnc e. The *solid-line* correspond to radius-mass relation for rocky planets such as the Earth, the *dotted lines* correspond to rocky cores which are surrounded by hydrogen/He gas envelopes that make up 1, 5, 10, and 20% of the total planetary mass, while the *dashed-lines* correspond to rocky cores which are surrounded by supercritical water layers

Table 3.8 Discovered super-Earths and hot Neptunes with known size and mass compared to relevant host star parameters such as spectral type, stellar mass, effective temperature T_{eff} and semi major axis d

Planet	R_{pl} (R_{Earth})	M_{pl} (M_{Earth})	Star-type	Star mass (M_{Sun})	T_{eff} (K)	d (AU)
GJ 1214b	~2.678	~6.55	M	0.153	2,949	0.014
GJ 436b	~4.3	~22.2	M2.5	0.452	3,684	0.02887
55 Cnc e	~2.0	~8.63	K01V-V	0.905	5,196	0.0156
CoRoT-7b	~1.58	~7.42	K0V	0.93	5,275	0.0172
HAT-P-11b	~4.58	~26.0	K4	0.81	4,780	0.053
Kepler-10b	~1.4	~4.56	G	0.895	5,672	0.01684
Kepler-4b	~3.87	~24.47	G0	1.223	5,857	0.0456
Kepler-11b	~1.97	~4.3	G	0.95	5,680	0.091
Kepler-11c	~3.15	~13.5	G	0.95	5,680	0.106
Kepler-11d	~3.43	~6.1	G	0.95	5,680	0.159
Kepler-11e	~4.52	~8.4	G	0.95	5,680	0.194
Kepler-11f	~2.61	~2.3	G	0.95	5,680	0.25

The order of the planets is selected after the spectral type and orbital location (status June 2012: http://www.exoplanet.eu/catalog.php)

1.87 g cm^{-3} [142] indicate substantial envelopes of light gases such as H and He or H_2O and H. Figure 3.23 shows the mass-radius relationship of discovered transiting super-Earths and hot Neptunes with known masses and in comparison with the Earth, Uranus, and Neptune. The solid black curve corresponds to models of planets with the Earth-like rock-iron composition. The higher dashed curve corresponds to 100%

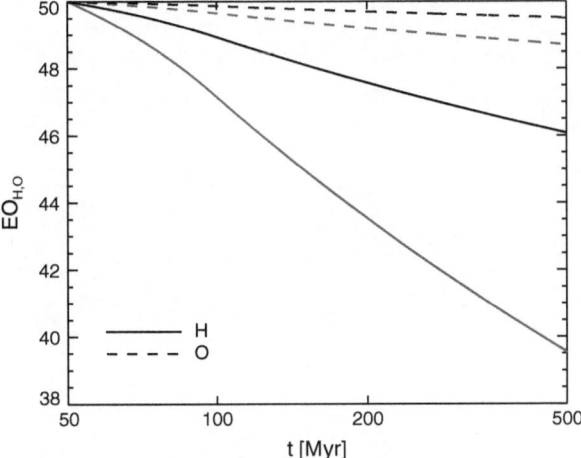

Fig. 3.24 Accumulation of abiotic oxygen around a super-Earth with a size of $1.5R_{Earth}$ and a mass of $5M_{Earth}$ at 1 AU inside the habitable zone of a Sun-like star, which can be produced by an outgassed steam atmosphere with a surface pressure of $\sim 10^4$ bar or an equivalent H_2O content of ~ 50 Earth oceans. The *black lines* correspond to a heating efficiency of 15 %, while the *gray lines* represent the atmospheric escape with a heating efficiency of 40 % (courtesy of P. Odert)

H. All other curves assume a H_2O or H_2/He envelope around a rocky iron core. The lower dashed-curve is 50 % H_2O by mass. The dotted curves are H_2/He gas envelopes that make up ~ 2, 6, 10, and 20 % of the total mass. For explaining the density of Kepler-11d, 11e, and 11f these planets require a dense H_2/He envelope, similar to Uranus and Neptune, while Kepler-11b and 11c may have H_2O and/or H_2/He gas envelopes [144]. The recent discovery of the transiting super-Earth 55 Cnc e with a radius of $\sim 2.08R_{Earth}$ and a mass of $\sim 7.8M_{Earth}$ with NASA's Spitzer space telescope in the 4.5 µm band points in the similar direction [145]. These authors report that similar to the above-mentioned low density super-Earths an Earth-like core which is surrounded either by a dense hydrogen/He gas envelope or a huge supercritical water envelope describes the observations at best. Table 3.8 shows the discovered super-Earths and hot Neptunes with known size and mass as well as their orbital distance and host star parameters such as spectral type, mass, and effective temperature T_{eff}. The discovery of these low mass hydrogen- and/or H_2O rich "rocky" planets is also in agreement with recent studies [84, 114, 148] and the discussions in this brief Monograph that one may expect that many super-Earths will not lose their initial hydrogen-rich protoatmospheres. The consequences of these findings for the evolution of the Earth-type class I habitats are very relevant. For instance, if the early Earth would have outgassed a steam atmosphere with a surface pressure $\gg 500$ bar, the planet might have had a problem in losing produced dense hydrogen and abiotic oxygen envelopes [84, 136]. Hydrogen and oxygen from an initially outgassed 1000 bar steam atmosphere during the EUV saturation phase of the young Sun, or a similar G-type star cannot be lost. In such a case, an Earth-type planet would retain more

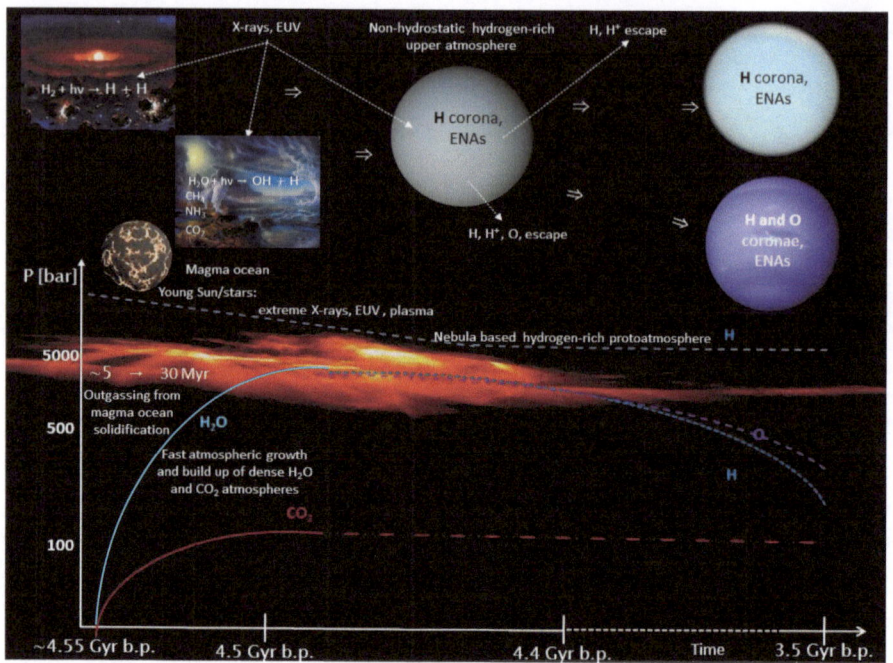

Fig. 3.25 Illustration of the formation, escape and evolution of nebula-based H-dominated protoatmospheres and/or magma ocean related catastrophically outgassed dense steam atmospheres from a super-Earth. The planet may capture and accumulate a huge amount of hydrogen, either from the nebula or from dissociated H_2O molecules. These dense hydrogen envelopes may not get lost completely from massive super-Earth's, especially when they orbit within habitable zones of Sun-like G stars or F stars. If the early planets were surrounded by dense steam atmospheres [114, 148] then a huge amount of abiotic oxygen which remains from dissociated water molecules could also populate the upper atmospheres

than 2.3 and 3.2 EO equivalent amount of hydrogen and oxygen, respectively. It is quite unlikely that such a huge amount of oxygen cannot be lost by non-thermal atmospheric escape processes [14, 149]. Super-Earths can outgas steam atmospheres of several 10^3 to several 10^4 bar surface pressure [114] may be good candidates for worlds with dense abiotic oxygen atmospheres.

Figure 3.24 demonstrates such a scenario for a super-Earth with a size of $1.5 R_{Earth}$ and a mass of $5 M_{Earth}$ at 1 AU inside the habitable zone of a Sun-like star, which outgassed a steam atmosphere with a surface pressure of $\sim 10^4$ bar or a H_2O content of ~ 50 EOs. One can see that the massive planet can not lose such a huge amount of water vapor. As illustrated in Fig. 3.25, as long as the water in the atmosphere remains in its gas phase the H_2O molecules will be dissociated by the high EUV flux of the young host star. Because of the high mass of the planet the hydrogen escape flux is too weak to drag the heavy oxygen away. Therefore, abiotic oxygen will accumulate in the upper atmosphere of such planets. From these results one can expect that there should be many planets, depending on their size, mass, orbital distance, as well as their host star's EUV flux environment, which may accumulate huge amount of abiotic oxygen.

The continuous discovery of low density super-Earths indicate that the atmospheric evolution of a planet which is dominated by a nitrogen atmosphere and a habitable environment such as on the Earth is strongly coupled to a complex interplay between its orbit location, the planetary formation process, water delivery and impact history, possible gas giants in the system, and the evolution of the host star's EUV and plasma environment.

By observing extended hydrogen exospheres and the velocities of the surrounding hydrogen atoms, due to the absorption of the host star's Lyman-α emission during exoplanet transits important insights into the evolutionary stage of the planet's atmosphere, its magnetic environment and the host star's plasma parameters could be drawn. Thus, the future detection and characterization of EUV-heated extended non-hydrostatic upper atmospheres around terrestrial exoplanets would confirm the discussed hypotheses. Because such observations will enhance our understanding how atmospheres of terrestrial planets evolve, in Chap. 4 a powerful method which allows a remote validation of these evolution scenarios, by future UV transit observations of terrestrial exoplanets with space telescopes such as the World Space Observatory-Ultra Violet (WSO-UV), will be discussed.

References

1. Léger, A., Selsis, F., Sotin, C., Guillot, T., Despois, D., Mawet, D., Ollivier, M., Labèque, A., Valette, C., Brachet, F., Chazelas, B., Lammer, H.: A new family of planets? "Ocean-planets". Icarus **169**, 499–504 (2004)
2. Selsis, F., Chazelas, B., Bordé, P., Ollivier, M., Brachet, F., Decaudin, M., Bouchy, F., Ehrenreich, D., Grießmeier, J.-M., Lammer, H., Sotin, C., Grasset, O., Moutou, C., Barge, P., Deleuil, M., Mawet, D., Despois, D., Kasting, J.F., Léger, A.: Could we identify hot ocean-planets with CoRoT, Kepler and doppler velocimetry? Icarus **191**, 453–468 (2007)
3. Tian, F., Kasting, J.F., Solomon, S.C.: Thermal escape of carbon from the early Martian atmosphere. Geophys. Res. Lett. **36**(2), CiteID L02205 (2009)
4. Lammer, H., Kasting, J.F., Chassefière, E., Johnson, R.E., Kulikov, Yu.N., Tian, F.: Atmospheric escape and evolution of terrestrial planets and satellites. Space Sci. Rev. **139**, 399–436 (2008)
5. Lammer, H., Kulikov, Yu.N., Lichtenegger, H.I.M.: Thermospheric X-ray and EUV heating by the young Sun on early Venus and Mars. Space Sci. Rev. **122**, 189–196 (2006)
6. Kulikov, Yu.N., Lammer, H., Lichtenegger, H.I.M., Penz, T., Breuer, D., Spohn, T., Lundin, R., Biernat, H.K.: A comparative study of the influence of the active young Sun on the early atmospheres of Earth, Venus and Mars. Space Sci. Rev. **129**, 207–243 (2007)
7. Lundin, R., Lammer, H., Ribas, I.: Planetary magnetic fields and solar forcing: implications for atmospheric evolution. Space Sci. Rev. **129**, 245–278 (2007)
8. Krauss, S., Fichtinger, B., Lammer, H., Hausleitner, W., Kulikov, Yu.N., Ribas, I., Shematovich, V.I., Bisikalo, D., Lichtenegger, H.I.M., Zaqarashvili, T.V., Khodachenko, M.L., Hanslmeier, A.: Solar flares as proxy for the young Sun: satellite observed thermosphere response to an X17.2 flare of Earth's upper atmosphere. Ann. Geophys. **30**, 1129–1141 (2012)
9. Kulikov, Yu.N., Lammer, H., Lichtenegger, H.I.M., Terada, N., Ribas, I., Kolb, C., Langmayr, D., Lundin, R., Guinan, E.F., Barabash, S., Biernat, H.K.: Atmospheric and water loss from early Venus. Planet. Space Sci. **54**, 1425–1444 (2006)

10. Tian, F., Kasting, J.F., Liu, H., Roble, R.G.: Hydrodynamic planetary thermosphere model: 1. The response of the Earth's thermosphere to extreme solar EUV conditions and the significance of adiabatic cooling. J. Geophys. Res. **113** (2008). doi:10.1029/2007JE002946

11. Tian, F., Solomon, S.C., Qian, L., Lei, J., Roble, R.G.: Hydrodynamic planetary thermosphere model: 2. Coupling of an electron transport/energy deposition model. J. Reophys. Res. **113**, E07005 (2008)

12. Tian, F.: Thermal escape from super Earth atmospheres in the habitable zones of M Stars. ApJ **703**, 905–909 (2009)

13. Chassefière, E.: Hydrodynamic escape of hydrogen from a hot water-rich atmosphere: the case of Venus. J. Geophys. Res. **101**, 26039–26056 (1996)

14. Chassefière, E.: Hydrodynamic escape of oxygen from primitive atmospheres: applications to the cases of Venus and Mars. Icarus **124**, 537–552 (1996)

15. Tian, F., Toon, O.B., Pavlov, A.A., De Sterck, H.: A hydrogen-rich early Earth atmosphere. Science **308**, 1014–1017 (2005)

16. Lichtenegger, H.I.M., Lammer, H., Grießmeier, J.-M., Kulikov, Yu.N., von Paris, P., Hausleit-ner, W., Krauss, S., Rauer, H.: Aeronomical evidence for higher CO_2 levels during Earth's Hadean epoch. Icarus **210**, 1–7 (2010)

17. Lammer, H., Bredehöft, J.H., Coustenis, A., Khodachenko, M.L., Kaltenegger, L., Grasset, O., Prieur, D., Raulin, F., Ehrenfreund, P., Yamauchi, M., Wahlund, J.-E., Grießmeier, J.-M., Stangl, G., Cockell, C.S., Kulikov, Yu.N., Grenfell, L., Rauer, H.: What makes a planet habitable? Astron. Astrophs. Rev. **17**, 181–249 (2009)

18. Lammer, H., Lichtenegger, H.I.M., Kulikov, Yu.N., Grießmeier, J.-M., Terada, N., Erkaev, N.V., Biernat, H.K., Khodachenko, M.L., Ribas, I., Penz, T., Selsis, F.: Coronal mass ejection (CME) activity of low mass M stars as an important factor for the habitability of terrestrial exoplanets. II. CME-induced ion pick up of Earth-like exoplanets in close-in habitable zones. Astrobiology **7**, 185–207 (2007)

19. Jeans, J.H.: The Dynamical Theory of Gases. Cambridge University Press, Cambridge (1925)

20. Bauer, S.J., Lammer, H.: Planetary Aeronomy: Atmosphere Environments in Planetary Systems. Springer, Berlin (2004)

21. Chamberlain, J.W.: Planetary coronae and atmospheric evaporation. Planet. Space. Sci. **11**, 901–996 (1963)

22. Pierrard, V.: Evaporation of hydrogen and helium atoms from the atmospheres of Earth and Mars. Planet. Space Sci. **51**, 319–327 (2003)

23. Hedelt, P., Ito, Y., Keller, H.U., Reulke, R., Wurz, P., Lammer, H., Rauer, H., Esposito, L.: Titan's atomic hydrogen corona. Icarus **210**, 424–435 (2010)

24. Öpik, E.J.: Selective escape of gases. Geophys. J. Roy. Astron. Soc. **7**, 490–509 (1963)

25. Penz, T., Erkaev, N.V., Kulikov, Yu.N., Langmayr, D., Lammer, H., Micela, G., Cecchi-Pestellini, C., Biernat, H.K., Selsis, F., Barge, P., Deleuil, M., Léger, A.: Mass loss from "Hot Jupiters"- Implications for CoRoT discoveries, Part II: long time thermal atmospheric evaporation modeling. Planet. Space Sci. **56**, 1260–1272 (2008)

26. Vidal-Madjar, A., Lecavelier des Etangs, A., Désert, J.M., Ballester, G.E., Ferlet, R., Hébrard, G., Mayor, M.: An extended upper atmosphere around the extrasolar planet HD209458 b. Nature **422**, 143–146 (2003)

27. Lammer, H., Lichtenegger, H.I.M., Kolb, C., Ribas, I., Guinan, E.F., Bauer, S.J.: Loss of water from Mars: implications for the oxidation of the soil. Icarus **165**, 9–25 (2003)

28. Erkaev, N.V., Penz, T., Lammer, H., Lichtenegger, H.I.M., Wurz, P., Biernat, H.K., Griess-meier, J.-M., Weiss, W.W.: Plasma and magnetic field parameters in the vicinity of short periodic giant exoplanets. ApJS **157**, 396–401 (2005)

29. Bauer, S.J.: Physics of Planetary Ionospheres. Springer, Berlin (1973)

30. Brace, L.H., Theis, R.F., Hoegy, W.R.: Plasma clouds above the ionopause of Venus and their implications. Planet. Space Sci. **30**, 29–37 (1982)

31. Elphic, R.C., Ershkovich, A.I.: On the stability of the ionopause of Venus. J. Geophys. Res. **89**, 997–1002 (1984)

32. Wolff, R.S., Goldstein, B.E., Yeates, C.M.: The onset and development of Kelvin-Helmholtz instability at the Venus ionopause. J. Geophys. Res. **85**, 7697–7707 (1980)
33. Penz, T., Erkaev, N.V., Biernat, H.K., Lammer, H., Amerstorfer, U.V., Gunell, H., Kallio, E., Barabash, S., Orsini, S., Milillo, A., Baumjohann, W.: Ion loss on Mars caused by the Kelvin-Helmholtz instability. Planet. Space Sci. **52**, 1157–1167 (2004)
34. Arshukova, I.L., Erkaev, N.V., Biernat, H.K., Vogl, D.F.: Interchange instability of the Venusian ionopause. Adv. Space Res. **33**, 182–186 (2004)
35. Lammer, H., Lichtenegger, H.I.M., Biernat, H.K., Erkaev, N.V., Arshukova, I.L., Kolb, C., Gunell, H., Lukyanov, A., Holmström, M., Barabash, S., Zhang, T.L., Baumjohann, W.: Loss of hydrogen and oxygen from the upper atmosphere of Venus. Planet. Space Sci. **54**, 1445–1456 (2006)
36. Chandrasekhar, S.: Hydrodynamic and Hydromagnetic Stability. Oxford University Press, New York (1961)
37. Pope, S.A., Balikhin, M.A., Zhang, T.L., Fedorov, A.O., Gedalin, M., Barabash, S.: Giant vortices lead to ion escape from Venus and re-distribution of plasma in the ionosphere. Geophys. Res. Lett. **36**, L07202 (2009)
38. Thomas, V.A., Winske, D.: Kinetic simulation of the Kelvin-Helmholtz instability at the Venus ionopause. Geophys. Res. Lett. **18**, 1943–1946 (1991)
39. Terada, N., Machida, S., Shinagawa, H.: Global hybrid simulation of the Kelvin-Helmholtz instability at the Venus ionopause. J. Geophys. Res. **107**, 1471–1490 (2002)
40. Amerstorfer, U.V., Erkaev, N.V., Langmayr, D., Biernat, H.K.: On Kelvin-Helmholtz instability due to the solar wind interaction with unmagnetized planets. Planet. Space Sci. **55**, 1811–1816 (2007)
41. Amerstorfer, U.V., Erkaev, N.V., Taubenschuss, U., Biernat, H.K.: Influence of a density increase on the evolution of the Kelvin-Helmholtz instability and vortices. Phys. Plasmas **17**, 072901 (2010)
42. Möstl, U.V., Erkaev, N.V., Zellinger, M., Lammer, H., Gröller, H., Biernat, H.K., Korovinskiy, D.: The Kelvin-Helmholtz instability at Venus: what is the unstable boundary? Icarus **216**, 476–484 (2011)
43. Pérez-de Tejada, H.: Plasma flow in the Mars magnetosphere. J. Geophys. Res. **92**, 4713–4718 (1987)
44. Pérez-de Tejada, H.: Momentum transport in the solar wind erosion of the Mars ionosphere. J. Geophys. Res. **103**, 31499–31508 (1998)
45. Lundin, R., Dubinin, E.M.: Phobos-2 results on the ionospheric plasma escape from Mars. Adv. Space Res. **12**, 255–263 (1992)
46. Hartle, R.E., Grebowsky, J.M.: Upward ion flow in ionospheric holes on Venus. J. Geophys. Res. **95**, 31–37 (1990)
47. Hartle, R.E., Grebowsky, J.M.: Light ion flow in the nightside ionosphere of Venus. J. Geophys. Res. **98**, 7437–7445 (1993)
48. Hartle, R.E., Donahue, T.M., Grebowsky, J.M., Mayr, H.G.: Hydrogen and deuterium in the thermosphere of Venus: solar cycle variations and escape. J. Geophys. Res. **101**, 4525–4538 (1996)
49. Lammer, H., Bauer, S.J.: A Mars magnetic field: constraints from molecular ion escape. J. Geophys. Res. **97**, 20925–20928 (1992)
50. Mc Elroy, M.B.: An evolving atmosphere. Science **175**, 443–445 (1972)
51. Nagy, A.F., Cravens, T.E., Yee, J.H., Stewart, A.I.F.: Hot oxygen atoms in the upper atmosphere of Venus. Geophys. Res. Lett. **8**, 629–632 (1981)
52. Ip, W.-H.: On a hot oxygen corona of Mars. Icarus **76**, 135–145 (1988)
53. Lammer, H., Bauer, S.J.: Non-thermal atmospheric escape from Mars and Titan. J. Geophys. Res. **96**, 1819–1825 (1991)
54. Fox, J.L., Hać, A.B.: Spectrum of hot O at the exobases of the terrestrial planets. J. Geophys. Res. **102**, 24005–24011 (1997)
55. Luhmann, J.: What do we really know about solar wind coupling? Adv. Space Res. **20**, 907–911 (1997)

56. Kim, J., Nagy, A.F., Fox, J.L., Cravens, T.E.: Solar cycle variability of hot oxygen atoms at Mars. J. Geophys. Res. **103**, 29339–29342 (1998)
57. Hodges Jr, R.R.: Distributions of hot oxygen for Venus and Mars. J. Geophys. Res. **105**, 6971–6981 (2000)
58. Lammer, H., Stumptner, W., Bauer, S.J.: Upper limits for the Martian exospheric number density during the planet B/Nozomi mission. Planet. Space Sci. **48**, 1473–1478 (2000)
59. Krestyanikova, M.A., Shematovich, V.I.: Stochastic models of hot planetary and satellite coronas: a photochemical source of hot oxygen in the upper atmosphere of Mars. Sol. Syst. Res. **39**, 2232 (2005)
60. Krestyanikova, M.A., Shematovich, V.I.: Stochastic models of hot planetary and satellite coronas: a hot oxygen corona of Mars. Sol. Syst. Res. **40**, 384–392 (2006)
61. Fox, J.L., Hać, A.B.: Photochemical escape of oxygen from Mars: a comparison of the exobase approximation to a Monte Carlo method. Icarus **204**, 527–544 (2009)
62. Chaufray, J.Y., Modolo, R., Leblanc, F., Chanteur, G., Johnson, R.E., Luhmann, J.G.: Mars solar wind interaction: formation of the Martian corona and atmospheric loss to space. J. Geophys. Res. **112**(E9), CiteID E09009 (2007)
63. Valeille, A., Combi, M.R., Tenishev, V., Bougher, S.W., Nagy, A.F.: A study of suprathermal oxygen atoms in Mars upper thermosphere and exosphere over the range of limiting conditions. Icarus **206**, 18–27 (2010)
64. Lichtenegger, H.I.M., Gröller, H., Lammer, H., Kulikov, Yu.N., Shematovich, V.I.: On the elusive hot oxygen corona of Venus. Geophys. Res. Lett. **36**, L10204 (2009)
65. Gröller, H., Shematovich, V.I., Lichtenegger, H.I.M., Lammer, H., Pfleger, M., Kulikov, Yu.N., Macher, W., Amerstorfer, U.V., Biernat, H.K.: Venus' atomic hot oxygen environment. J. Geophys. Res. **115**, E12017 (2010)
66. Gröller, H., Lammer, H., Lichtenegger, H.I.M., Pfleger, M., Dutuit, O., Shematovich, V.I., Kulikov, Yu.N., Biernat, H.K.: Hot oxygen atoms in the Venus nightside exosphere. Geophys. Res. Lett. **39**, L03202 (2012)
67. Gurwell, M.A., Yung, Y.L.: Fractionation of hydrogen and deuterium on Venus due to collisional ejection. Planet. Space Sci. **41**, 91–101 (1993)
68. Balakrishnan, N., Kharchenko, V., Dalgarno, A.: Slowing of energetic O(^{3}P) atoms in collisions with N_2. J. Geophys. Res. **103**, 23392–23398 (1998)
69. Balakrishnan, N., Kharchenko, V., Dalgarno, A.: Quantum mechanical and semiclassical studies of N^+-N_2 collisions and their application to thermalization of fast N atoms. J. Chem. Phys. **108**, 943–949 (1998)
70. Kharchenko, V., Dalgarno, A., Zygelman, B., Yee, J.H.: Energy transfer in collisions of oxygen atoms in the terrestrial atmosphere. J. Geophys. Res. **103**, 24899–24906 (2000)
71. Jakosky, B.M., Pepin, R.O., Johnson, R.E., Fox, J.L.: Mars atmospheric loss and isotopic fractionation by solar-wind-induced sputtering and photochemical escape. Icarus **111**, 271–288 (1994)
72. Leblanc, F., Johnson, R.E.: Role of molecular species in pick up ion sputtering of the Martian atmosphere. J. Geophys. Res. **107**, 1–6 (2002)
73. Johnson, R.E.: Energetic Charged Particle Interactions with Atmospheres and Surfaces. Springer, Berlin (1990)
74. Lammer, H., Bauer, S.J.: Atmospheric mass loss from Titan by sputtering. Planet. Space Sci. **41**, 657–663 (1993)
75. Luhmann, J.G., Kozyra, J.U.: Dayside pickup oxygen ion precipitation at Venus and Mars: spatial distributions, energy deposition and consequences. J. Geophys. Res. **96**, 5457–5467 (1991)
76. Terada, N., Kulikov, Yu.N., Lammer, H., Lichtenegger, H.I.M., Tanaka, T., Shinagawa, H., Zhang, T.-L.: Atmosphere and water loss form early Mars under extreme solar wind and extreme ultraviolet conditions. Astrobiology **9**, 55–70 (2009)
77. Scalo, J., Kaltenegger, L., Segura, A.G., Fridlund, M., Ribas, I.: Kulikov, Yu.N., Grenfell, J.L., Rauer, H., Odert, P., Leitzinger, M., Selsis, F., Khodachenko, M.L., Eiroa, C., Kasting, J., Lammer, H.: M stars as targets for terrestrial exoplanet searches and biosignature detection. Astrobiology **7**, 85–166 (2007)

78. Yelle, R.V.: Aeronomy of extra-solar giant planets at small orbital distances. Icarus **170**, 167–179 (2004)
79. Koskinen, T.T., Yelle, R.V., Lavvas, P., Lewis, N.K.: Characterizing the thermosphere of HD209458 b with UV tranist observations. ApJ **723**, 116–128 (2010)
80. Zahnle, K.J., Kasting, J.F., Pollack, J.B.: Evolution of a steam atmosphere during Earth's accretion. Icarus **74**, 62–97 (1988)
81. Sekiya, M., Nakazawa, K., Hayashi, C.: Dissipation of the primordial terrestrial atmosphere due to irradiation of the solar EUV. Prog. Theor. Phys. **64**, 1968–1985 (1980)
82. Sekiya, M., Nakazawa, K., Hayashi, C.: Dissipation of the rare gases contained in the primordial Earth's atmosphere. Earth Planet. Sci. Lett. **50**, 197–201 (1980)
83. Sekiya, M., Hayashi, C., Nakazawa, K.: Dissipation of the primordial terrestrial atmosphere due to irradiation of the solar far-UV during T-Tauri stage. Prog. Theor. Phys. **66**, 1301–1316 (1981)
84. Lammer, H., Kislyakova, K.G., Odert, P., Leitzinger, M., Schwarz, R., Pilat-Lohinger, E., Kulikov, Yu.N., Khodachenko, M.L., Güdel, M., Hanslmeier, A.: Pathways to Earth-like atmospheres: extreme ultraviolet (EUV)-powered escape of hydrogen-rich protoatmospheres. Orig. Life Evol. Biosph. **41**, 503–522 (2012)
85. Zahnle, K.J., Walker, J.C.G.: The evolution of solar ultraviolet luminosity. Rev. Geophys. **20**, 280–292 (1982)
86. Güdel, M., Guinan, E.F., Skinner, S.L.: The X-ray Sun in Time: a study of the long-term evolution of coronae of solar-type stars. ApJ **483**, 947–960 (1997)
87. Ribas, I., Guinan, E.F., Güdel, M., Audard, M.: Evolution of the solar activity over time and effects on planetary atmospheres. I. High-energy irradiances (1–1700 Å). ApJ **622**, 680–694 (2005)
88. Güdel, M.: The Sun in time: activity and environment. Liv. Rev. Solar Phys. **4**(3), 1–137 (2007)
89. Hartmann, L., Kenyon, S.J.: High spectral resolution infrared observations of V1057 Cygni. ApJ **322**, 393–398 (1987)
90. Hartmann, L., Kenyon, S.J.: The FU Orionis phenomenon. Ann. Rev. Astron. Astrophys. **34**, 207–240 (1996)
91. Lammer, H., Stumptner, W., Molina-Cuberos, G.J., Bauer, S.J., Owen, T.: Nitrogen isotope fractionation and its consequence for Titan's atmospheric evolution. Planet. Space Sci. **48**, 529–543 (2000)
92. Feigelson, E.D., Montmerle, T.: High-energy processes in young stellar objects. Ann. Rev. Astron. Astrophys. **37**, 363–408 (1999)
93. Checchi-Pestellini, C., Ciaravella, A., Micela, G.: Stellar X-ray heating of planetary atmospheres. A&AL **458**, L13–L16 (2006)
94. Owen, J., Jackson, A.: Planetary evaporation by UV and X-ray radiation: basic hydrodynamics. Mon. Not. R. Astron. Soc. (2012) (accepted)
95. Lammer, H., Odert, P., Leitzinger, M., Khodachenko, M.L., Panchenko, M., Kulikov, Yu.N., Zhang, T.L., Lichtenegger, H.I.M., Erkaev, N.V., Wuchterl, G., Micela, G., Penz, A., Biernat, H.K., Weingrill, J., Steller, M., Ottacher, H., Hasiba, J., Hanslmeier, A.: Determining the mass loss limit for close-in exoplanets: what can we learn from transit observations? A&A **506**, 399–410 (2009)
96. Waite Jr, J.H., Cravens, T.E., Kozyra, J., Nagy, A.F., Atreya, S.K., Chen, R.H.: Electron precipitation and related aeronomy of the Jovian thermosphere and ionosphere. J. Geophys. Res. **88**, 6143–6163 (1983)
97. Murray-Clay, R.A., Chiang, E.I., Murray, N.: Atmospheric escape from hot Jupiters. ApJ **693**, 23–42 (2009)
98. Leitzinger, M., Odert, P., Kulikov, Yu.N., Lammer, H., Wuchterl, G., Penz, T., Guarcello, M.G., Micela, G., Khodachenko, M.L., Weingrill, J., Hanslmeier, A., Biernat, H.K., Schneider, J.: Could CoRoT-7b and Kepler-10b be remnants of evaporated gas or ice giants? Planet. Space Sci. **59**, 1472–1481 (2011)
99. Hunten, D.M.: Atmospheric evolution of the terrestrial planets. Science **259**, 915–920 (1993)

100. Elkins-Tanton, L.T.: Linked magma ocean solidification and atmospheric growth for Earth and Mars. Earth Planet. Sci. Lett. **271**, 181–191 (2008)
101. Liu, L.-G.: The inception of the oceans and CO_2-atmosphere in the early history of the Earth. Earth Planet. Sci. Lett. **227**, 179–184 (2004)
102. Allègre, C.J., Hofmann, A.W., O'Nions, R.K.: The argon constraints on mantle structure. Geophys. Res. Lett. **23**, 3555–3557 (1996)
103. Touboul, M., Kleine, T., Bourdon, B., Palme, H., Wieler, R.: Late formation and prolonged differentiation of the Moon inferred from W isotopes in lunar metals. Nature **450**, 1206–1209 (2007)
104. Mizuno, H.: Formation of the giant planets. Prog. Theor. Phys. **64**, 544–557 (1980)
105. Hayashi, C., Nakazawa, K., Mizuno, H.: Earth's melting due to the blanketing effect of the primordial dense atmosphere. Earth Planet. Sci. Lett. **43**, 22–28 (1979)
106. Matsui, T., Abe, Y.: Impact-induced atmospheres and oceans on Earth and Venus. Nature **322**, 526–528 (1986)
107. Albarède, F., Blichert-Toft, J.: The split fate of the early Earth, Mars, Venus and Moon. CR Geosci. **339**, 917–927 (2007)
108. Abe, Y.: Thermal and chemical evolution of the terrestrial magma ocean. Phys. Earth Planet. Int. **100**, 27–39 (1997)
109. Zahnle, K.J., Kasting, J.F.: Mass fractionation during transonic escape and implications for loss of water from Mars and Venus. Icarus **68**, 462–480 (1986)
110. Hunten, D.M., Pepin, R.O., Walker, J.C.G.: Mass fractionation in hydrodynamic escape. Icarus **69**, 532–549 (1987)
111. Lammer, H., Chassefière, E., Karatekin, Ö, Morschhauser, A., Niles, P.B., Mousis, O., Grott, M., Gröller, H., Hauber, E., Pham, L.B.S.: Outgassing history and escape of the martian atmosphere and water inventory. Space Sci. Rev. (2012) (accepted)
112. Wood, B.E., Müller, H.-R., Zank, G., Linsky, J.L.: Measured mass loss rates of solar-like stars as a function of age and activity. ApJ **574**, 412–425 (2002)
113. Gender, H., Aber, Y.: Survival of a proto-atmosphere through the stage of giant impacts: the mechanical aspects. Icarus **164**, 149–162 (2003)
114. Elkins-Tanton, L.T.: Formation of water ocean on rocky planets. Astrophys. Space Sci. **332**, 359–364 (2011)
115. Kasting, J.F., Pollack, J.B., Crisp, D.: Effects of high CO_2 levels on surface temperature and atmospheric oxidation state of the early Earth. J. Atmos. Chem. **1**, 403–428 (1984)
116. Kempe, S., Degens, E.T.: An early Soda ocean? Chem. Geol. **53**, 95–108 (1985)
117. Lunine, J.I., Chambers, J., Morbidelli, A., Leshin, L.A.: The origin of water on Mars. Icarus **165**, 1–8 (2003)
118. Horner, J., Mousis, O., Petit, J.-M., Jones, B.-W.: Differences between the impact regimes of the terrestrial planets: implications for primordial D:H ratios. Planet. Space Sci. **57**, 1338–1345 (2009)
119. Chassefière, E., Leblanc, F.: Constraining methane release due to serpentinzation by the observed D/H ratio on Mars. Earth Planet. Sci. Lett. **310**, 262–271 (2011)
120. Lammer, H., Kolb, C., Penz, T., Amerstorfer, U.V., Biernat, H.K., Bodiselitsch, B.: Estimation of the past and present Martian water-ice reservoirs by isotopic constraints on exchange between the atmosphere and the surface. Int. J. Astrobiol. **2**, 195–202 (2003b)
121. Barabash, S., Fedorov, A., Lundin, R., Sauvaud, J.-A.: Martian atmospheric erosion rates. Science **315**, 501–503 (2007)
122. Ma, Y.-J., Nagy, A.F.: Ion escape fluxes from Mars. Geophys. Res. Lett. **34**, L08201 (2007)
123. Modolo, R., Chanteur, G.M., Dubinin, E., Matthews, A.P.: Influence of the solar EUV flux on the Martian plasma environment. Ann. Geophys. **23**, 1–12 (2005)
124. Chassefière, E., Leblanc, F., Langlais, B.: The combined effects of escape and magnetic field histories at Mars. Planet. Space Sci. **55**, 343–357 (2007)
125. Manning, C.V., Ma, Y., Brain, D.A., McKay, C.P., Zahnle, K.J.: Parametric analysis of modeled ion escape from Mars. Icarus **212**, 131–137 (2010)

126. Chappell, C.R., Moore, T.E., Waite Jr, J.H.: The ionosphere as a fully adequate source of plasma for the earth's magnetosphere. J. Geophys. Res. **92**, 5896–5910 (1987)
127. Moore, T.E., Lundin, R., Alcayde, D., Andre, M., Ganguli, S.B., Temerin, M., Yau, A.: Source processes in the high-latitude ionosphere. Space Sci. Rev. **88**, 7–84 (1999)
128. Seki, K., Elphic, R.C., Hirahara, M., Terasawa, T., Mukai, T.: On atmospheric loss of oxygen ions from Earth through magnetospheric processes. Science **291**, 1939–1941 (2001)
129. Yau, A.W., André, M.: Sources of ion outflow in the high latitude ionosphere. Space Sci. Rev. **37**, 1–25 (1997)
130. Wei, Y., Fraenz, M., Dubinin, E., Woch, J., Lühr, H., Wan, W., Zong, Q.-G., Zhang, T.-L., Pu, Z.Y., Fu, S.Y., Barabash, S., Lundin, R., Dandouras, I.: Enhanced atmospheric oxygen outflow on Earth and Mars driven by a corotating interaction region. J. Geophys. Res. **117**, A03208 (2012)
131. Roble, R.G., Rodley, E.C., Dickinson, R.E.: On the global mean structure of the thermosphere. J. Geophys. Res. **92**, 8745–8758 (1987)
132. Smithtro, C.G., Sojka, J.J.: A new global average model of the coupled thermosphere and ionosphere. J. Geophys. Res. **110**, A08305 (2005a)
133. Smithtro, C.G., Sojka, J.J.: Behavior of the ionosphere and thermosphere subject to extreme solar cycle conditions. J. Geophys. Res. **110**, A08306 (2005b)
134. Trigo-Rodriguez, J.M., Javier Martín-Torres, F.: Clues on the importance of comets in the origin and evolution of the atmospheres of Titan and Earth. Planet. Space Sci. **60**, 3–9 (2012)
135. Lammer, H., Kislyakova, K.G., Güdel, M., Holmström, M., Erkaev, N.V., Odert, P., Khodachenko, M.L.: Stability of Earth-like N_2 atmospheres: implications for habitability. In: Muller, C., Nixon, C.A., Raulin, F., Trigo-Rodriguez, J.M. (eds.) Nitrogen. Springer, Heidelberg (2012) (in press)
136. Lammer, H., Güdel, M., Kulikov, Yu.N., Ribas, I., Zaqarashvili, T.V., Khodachenko, M.L., Kislyakova, K.G., Gröller, H., Odert, P., Leitzinger, M., Fichtinger, B., Krauss, S., Hausleitner, W., Holmström, M., Sanz-Forcada, J., Lichtenegger, H.I.M., Hanslmeier, A., Shematovich, V.I., Bisikalo, D., Rauer, H., Fridlund, M.: Variability of solar/stellar activity and magnetic field and its influence on planetary atmosphere evolution. Earth Planets Space **63**, 179–199 (2012)
137. Rivera, E.J., Lissauer, J.J., Butler, R.P., Marcy, G.W., Vogt, S.S., Fischer, D.A., Brown, T.M., Laughlin, G.H., Gregory, W.: A $\sim 7.5 M_{Earth}$ planet orbiting the nearby star, GJ 876. ApJ **634**, 625–640 (2005)
138. Gould, A., Udalski, A., An, D., Bennett, D.P., Zhou, A.-Y., Dong, S., Rattenbury, N.J., Gaudi, B.S., Yock, P.C.M., Bond, I.A., Christie, G.W., Horne, K., Anderson, J., Stanek, K.Z., DePoy, D.L., Han, C., McCormick, J., Park, B.-G., Pogge, R.W., Poindexter, S.D., Soszyński, I., Szymański, M.K., Kubiak, M., Pietrzyński, G., Szewczyk, O., Wyrzykowski, L., Ulaczyk, K., Paczyński, B., Bramich, D.M., Snodgrass, C., Steele, I.A., Burgdorf, M.J., Bode, M.F., Botzler, C.S., Mao, S., Swaving, S.C.: Microlens OGLE-2005-BLG-169 implies that cool Neptune-like planets are common. ApJ **644**, L37–L40 (2006)
139. Lovis, C., Mayor, M., Pepe, F., Alibert, Y., Benz, W., Bouchy, F., Correia, A.C.M., Laskar, J., Mordasini, C., Queloz, D., Santos, N.C., Udry, S., Bertaux, J.-L., Sivan, J.-P.: An extrasolar planetary system with three Neptune-mass planets. Nature **441**, 305–309 (2006)
140. Beust, H., Bonfils, X., Delfosse, X., Udry, S.: Dynamical evolution of the Gliese 581 planetary system. A&A **479**, 277–282 (2008)
141. Léger, A., Rouan, D., Schneider, J., Barge, P., Fridlund, F., CoRoT Team: Transiting exoplanets from the CoRoT space mission VIII. CoRoT-7b: The first super-Earth with measured radius. A&A **506**, 287–302 (2009)
142. Charbonneau, D., Berta, Z.K., Irwin, J., Burke, C.J., Nutzman, P., Buchhave, L.A., Lovis, C., Bonfils, X., Latham, D.W., Udry, S., Murray-Clay, R.A., Holman, M.J., Falco, E.E., Winn, J.N., Queloz, D., Pepe, F., Mayor, M., Delfose, X., Forveille, T.: A super-Earth transiting a nearby low-mass star. Nature **462**, 891–894 (2009)
143. Batalha, N.M., Borucki, W.J., Bryson, S.T., Buchhave, L.A., Caldwell, D.A., Kepler Team: Kepler's first rocky planet: Kepler-10b. ApJ **729**, article id. 27 (2011)

144. Lissauer, J.J., Kepler Team: A closely packed system of low-mass, low-density planets transiting Kepler-11. Nature **470**, 53–58 (2011)
145. Demory, B.-O., Gillon, M., Deming, D., Valencia, D., Seager, S., Benneke, B., Lovis, C., Cubillos, P., Harrington, J., Stevenson, K.B., Mayor, M., Pepe, F., Queloz, D., Ségransan, D., Udry, S.: Detection of a transit of the super-Earth 55 Cnc e with warm Spitzer. A&A, 553, id.A114 (2011) (Submitted)
146. Bonfils, X., Delfosse, X., Udry, S., Forveille, T., Mayor, M., Perrier, C., Bouchy, F., Gillon, M., Lovis, C., Pepe, F., Queloz, D., Santos, N.C., Ségransan, D., Bertaux, J.-L.: The HARPS search for southern extra-solar planets XXXI. The M-dwarf sample. A&A, 2011arXiv1111.5019B (2011) (Submitted)
147. Borucki, B., Koch, D.G., Basri, G., Batalha, N., Brown, T.M., Kepler Team: Characteristic of Kepler planetary candidates based on the first data set. ApJ **728**, 117 (20pp) (2011)
148. Elkins-Tanton, L., Seager, S.: Ranges of atmospheric mass and composition of super-Earth exoplanets. ApJ **685**, 1237–1246 (2008)
149. Kasting, J.F.: O_2 concentrations in dense primitive atmospheres: commentary. Planet. Space Sci. **43**, 11–13 (1995)

Chapter 4
Observational Tests of Atmosphere Evolution Hypotheses

Spacecraft observations of hydrogen Energetic Neutral Atoms (ENAs) and the application of advanced numerical models developed to a relevant remote-sensing technique in planetary and space science. Hydrogen ENAs are produced whenever a solar- or stellar wind proton interacts via charge exchange with a neutral atom from an upper atmosphere so that their signals contain the information from the structure of the upper atmosphere and its neutral gas density, as well as that from the plasma environment around a planetary obstacle. By combining these observations with theoretical models of the solar wind plasma flow and its interaction with the upper atmospheres of planetary bodies and comets can be analyzed and studied to a great accuracy. From comparative studies between Solar System planets and exoplanets one can expect that similar processes will also occur within their environments. Exoplanets which are in orbit locations closer to their host stars, or within close-in habitable zones of active dwarf stars can be seen as proxies of Solar System planets during the time of the young Sun period.

4.1 The Role of Energetic Neutral Atoms

Magnetic planets such as the Earth contain large regions which are filled with trapped electrons and ions with energies ranging from thermal plasma with energies in the order of a few eV to plasma in the high-energy radiation belts with energies $\geq 100\,\text{keV}$. This region, the magnetosphere is bounded by the magnetopause where the pressure balance between the geomagnetic field and the solar wind plasma can be balanced.

It took, a lot of research leading to the realization that energetic ions trapped in the magnetosphere that are neutralized by charge exchange with exospheric particles can escape their entrapment and therefore be detected far from their source [1, 2]. From this discovery plasma physicists become aware that these newly generated ENAs can also be used for an imaging technique, which has finally provided accurate insights into the global plasma dynamics around various planets in the Solar System.

H. Lammer, *Origin and Evolution of Planetary Atmospheres,*
SpringerBriefs in Astronomy, DOI: 10.1007/978-3-642-32087-3_4,
© The Author(s) 2013

During the charge exchange, a singly charged energetic ion picks up an electron from a neutral atom or molecule of the exosphere, so the ion is thereby neutralized and becomes an ENA

$$H_{sw}^{+} + H_{pl} \rightarrow H_{sw}^{ENA} + H_{pl}^{+}. \qquad (4.1)$$

The ENA which originated from a solar wind proton is now a neutral atom, its motion is no longer governed and influenced by magnetic and electric fields. If the ENA has a sufficient energy, it is also not affected by the gravitational field, so that the atom propagates in a straight trajectory, similar as photons. The flux of ENAs F_{ENA} at energy E in units of ENA number per unit area, per solid angle, per unit time, and per unit energy can thus be written as a line-of-sight integral over the ion flux F_{ion}

$$F_{ENA}(E) = \int_{0}^{\infty} F_{ion}(E, z) n(z) \sigma_{ch}(z) dz, \qquad (4.2)$$

where the number density of the cooler neutral particles n in which the plasma is submerged is multiplied by the charge exchange cross section σ_{ch}. On Earth, ENA images of the inner magnetosphere are used to obtain global information about the ring current morphology, dynamics, and composition.

ENA imaging is also used for gathering information on the plasma environment beyond the Earth's magnetopause and its time variation related to solar activity and the coupling between solar-terrestrial relations. The shocked solar wind in the Earth's magnetosheath becomes nearly stationary at the subsolar magnetopause. At this location, solar wind protons are neutralized by charge exchange with neutral hydrogen atoms at the extreme limits of the Earth's exosphere.

The newly produced ENAs propagate away from the subsolar region in nearly all directions. Analyzed simultaneous observations of ENAs from the Interstellar Boundary Explorer (IBEX) spacecraft and observed proton distributions in the magnetosheath from the Cluster satellites have been used to quantify the charge exchange process near the Earth's subsolar magnetopause standoff distance. By combining such observations, the ratio of ENAs to the shocked solar wind flux is $\sim 10^{-4}$ and the exospheric density at $\geq 9 r_{Earth}$ upstream from the Earth's surface is estimated to be $\sim 8 \, cm^{-3}$ [3].

Besides the Earth, ENAs have also been observed around Mars and Venus [4–8] as well as on Jupiter and Saturn. As shown in Fig. 4.1, ENAs and suprathermal atoms played also an important role in the estimation of exosphere temperatures based by misinterpretations of early Lyman-α observations around Mars and Venus [9]. The first observations of extended atomic hydrogen exospheres around Mars (Mariner 6 and 7: [10]) and Venus (Mariner 5: [11]) were inferred from remote Lyman-α observations. Exosphere temperatures of $\sim 350 \pm 100 \, K$ for Mars [10] and ~ 700–$1,000 \, K$ for Venus [11] were estimated and interpreted as escaping planetary hydrogen atoms. Temperatures inferred from mass spectrometer and aerobreaking data [12] yielded much lower values of ~ 220 and $\sim 300 \, K$ for Mars and Venus, respectively. ENAs which originate via charge exchange of solar wind protons with exospheric neutral

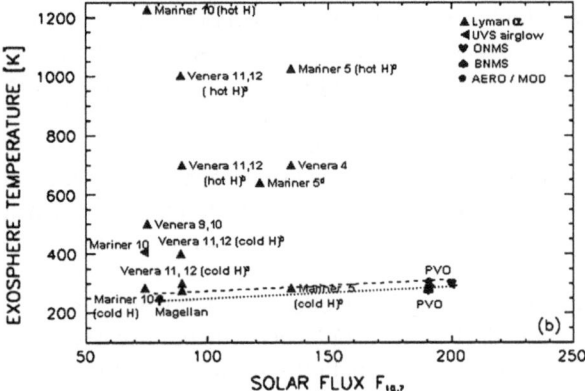

Fig. 4.1 Daytime Venus exosphere temperature estimations based on the Lyman-α, UVS airglow data analysis during the solar cycle. Exobase temperatures inferred from NASA's Magellan and PVO spacecraft from aerobraking maneuvers and mass spectrometry measurements are also shown. The *two dashed lines* correspond to theoretical models (after [9])

particles around Mars and Venus can influence the exospheric temperature inferred from Lyman-α UV data together with photochemically produced suprathermal H atoms [13]. On its travel to Saturn, the Cassini spacecraft imaged ENAs during its flyby at the Jupiter magnetosphere [14]. The images also revealed charge exchange processes between energetic ions and the neutral gas torus which is produced by Jupiter's icy moon Europa. When Cassini arrived in the Saturn system, ENAs were also imaged and it was found that Saturn's large moon Titan is a strong ENA source due to the interaction of its upper atmosphere with the plasma flow of Saturn's magnetosphere [15, 16].

Since the past decade the study of the production of ENAs and their topology around bodies which are immersed in a plasma flow was even extended to the Sun. When the solar wind flows away from the Sun it carries with it the solar magnetic field and produces a magnetic cavity in the local ISM. This cavity is called the heliosphere which encompasses the whole Solar System and likely extends outward from the Sun up to $\sim$150–200 AU. Because the ISM consists mainly of neutral and ionized hydrogen and helium atoms, the Sun's heliosphere is the area which is dominated by its magnetized plasma, including the supersonic solar wind, which becomes subsonic when it meets the local ISM at the termination shock at $\sim$100 AU. The heliosheath consists of the shocked solar wind plasma and extends another $\sim$50–100 AU outward from the termination shock.

The only possible global measurements of the heliosheath are remote techniques, such as ENA imaging [17]. The protons of the hot and relatively dense heliosheath charge exchange with the neutral gas of the local ISM, ENAs are produced from all areas of the heliosheath. From these emissions the real nature of the solar-wind/ISM interaction can be studied. Although the neutral gas densities are low, the enormous thickness of the heliosheath makes it the dominant ENA source. Similar techniques in combination with Hubble Space Telescope (HST) Lyman-α observations have also been used for mass loss observations of nearby stars by observing the interaction regions carved out by the collisions between stellar winds and the ISM [18, 19].

Thus, the observations of ENAs around Solar System planets, the Sun, and astrospheres of other stars can also be seen as a confirmation that huge ENA-formation zones should also be produced in the stellar wind interaction region around the planetary obstacles of hot Jupiters where the exosphere densities are several orders of magnitude larger compared to the observed ones at Venus, the Earth, or Mars [20–22].

In the following sections, the production of stellar wind-related ENAs and hydrogen exospheres around close-in exoplanets will be discussed and shown how a detailed analysis of Lyman-α attenuation spectra obtained from transiting exoplanets can be used for studying of the upper atmosphere structure, the planet's magnetosphere and stellar wind properties.

4.2 UV-Transit Observations and Modeling as a Remote Sensing Tool for Analyzing Stellar-Exoplanet Interaction Processes

Nearly two decades after the discovery of 51 Peg b, the first Jupiter-type gas giant outside our Solar System, about 800 exoplanets have been discovered. Although most exoplanets are detected with the radial velocity method only 4 years after the observation of 51 Peg b, the first transiting hot Jupiter, HD 209458b with a size of $\sim 1.38 R_{Jup}$ and mass of $\sim 0.64 M_{Jup}$ was observed [23]. The detection of Jupiter-type planets at orbital distances ≤ 0.05 AU soon opened questions regarding their upper atmosphere structure, the extreme stellar plasma interaction and the stability against escape of atmospheric gas [24, 25]. An analytical model of exoplanetary upper atmospheres which is based on an approximate solution of the heat balance equation in thermospheres was soon after the discovery of HD 209458b developed [26]. It was found that a hydrogen-rich thermosphere of a Jupiter-type planet in close orbital distance will be heated to several thousand Kelvin so that similar as in the theoretical studies which have been discussed in previous sections, hydrostatic conditions will not be valid anymore and the thermosphere will dynamically expand.

Transit observations of HD 209458b with the HST/Space Telescope Imaging Spectrograph (STIS) instrument discovered a $15 \pm 4\%$ intensity drop in the stellar Lyman-α line in the high velocity part of the spectra which was explained as evaporating neutral H atoms which are accelerated by the stellar radiation pressure [27]. Since the past years the estimation of the stellar Lyman-α absorption rate has also been carried out by more researchers [28] and subsequent observations at low spectral resolution with the HST STIS/Advanced Camera for Surveys (ACS) have now confirmed that the transit depth in Lyman-α is most likely $\sim 9\%$ larger than the transit depth due to the planetary disk alone [29]. Recently, another observation of an extended upper atmosphere due to Lyman-α absorption during the transits of the short periodic gas giant, HD 189733b was reported [30]. Due to natural and thermal broadening of the spectral lines and large column densities along the lines-of-sight through the atmosphere, the optical depth in the wings of the line profile can be significant even in the absence of actual bulk flows towards or away from the observer. By assuming column contents $\geq 10^{22}\,\mathrm{m}^{-2}$, broadening alone may explain the transit observations in the Lyman-α light curve at HD 209458b, this process may have a problem to explain the absorption data of $5.05 \pm 0.75\,\%$ [30, 31] at HD 189733b. Because this hot Jupiter orbits a more active K-type star at a closer orbital distance $\sim 80\,\%$ of the neutral H atoms in the thermosphere are ionized [31]. One obtains column contents which are much lower ($\ll 10^{22}\,\mathrm{m}^{-2}$) than the values used in the empirical thermosphere models of HD 209458b.

Applied hydrodynamic models [32–35] and empirical thermosphere models [36] also indicate that close-in exoplanets experience extreme EUV heating, atmospheric expansion, and outflow up to their Roche lobes with mass loss rates $\sim 10^{10}$–$5 \times 10^{10}\,\mathrm{g}\,\mathrm{s}^{-1}$ [37]. This is also supported by independent HST/STIS/Cosmic Origins Spectrograph (COS) observations where carbon, oxygen, and Si were observed beyond the Roche lobe of HD 209458b [38, 39] and various metals around the Roche lobe of WASP-12b [40]. As discussed in Chap. 3 it is possible that heavy atmospheric species, which are located in the lower thermosphere, can be dragged upward by the dynamically expanding hydrogen thermosphere. The exosphere-magnetosphere environment of hydrogen-rich gas giants in orbit locations ≤ 0.05 AU should also be strongly affected by the production of non-thermal neutral H atoms, and ENAs as well as non-thermal ion escape processes [20–22, 31, 41–46].

By applying a coupled plasma flow and exosphere Monte Carlo model to the stellar wind plasma interaction with HD 209458b one can calculate the ENA production near the planet [20, 21]. One can assume that charge exchange between stellar wind protons and the hydrodynamically outward flowing planetary neutral hydrogen atoms takes place in the undisturbed stellar wind, which flows radially away from the host star. To simulate the ENA production one needs a stellar-wind flow and neutral atmosphere model. The combination of these two models together with the cross-section σ_{ch}^{j} for charge exchange between stellar-wind protons n_{sw} and planetary neutral atoms n_{n} of species j allow to compute the ENA production Q_{ENA} rate within a volume around the planet

$$Q_{\mathrm{ENA}} = n_{\mathrm{sw}} v_{\mathrm{sw}} \sum_{j} \sigma_{\mathrm{ch}}^{j} n_{\mathrm{n}}^{j}, \qquad (4.3)$$

with the stellar wind velocity v_{sw}.

Because HD 209458 is a solar-like star with an age of ~ 4 Gyr solar parameters could be scaled from 1 AU to the orbit location of HD 209458b at 0.047 AU. The density of the outward flowing atmospheric hydrogen atoms can be calculated with hydrodynamical models [32, 33, 35] or by using empirical model results [36]. In these studies the blue part of the velocity spectrum (-200 to $-50\,\mathrm{km}\,\mathrm{s}^{-1}$) even by different assumed stellar wind velocities could be fitted well ($v_{\mathrm{sw}} = 50\,\mathrm{km}\,\mathrm{s}^{-1}$: [20];

$v_{sw} = 450$ km s^{-1}: [21]). Some Lyman-α absorption in the red part of the velocity spectrum ($+75$ km s^{-1}) was obtained but the fits were very poor compared to the blue part. So far the absorption of neutral hydrogen atoms which move with velocities between -30 and 110 km s^{-1} toward the star cannot be explained by the radiation pressure hypothesis [27] nor very well with the published ENA studies [20, 21].

The before discussed ENA studies assumed similar conically magnetopause obstacles with a constant magnetopause distance at the planet's terminator $R_t = 20R_{pl}$ and only the sub-stellar standoff distance R_s was varied. Moreover, these previous studies did not include the possible effect of line broadening. During the transit of an exoplanet the Lyman-α line emissions from the star will be attenuated by scattering of the photons by the hydrogen atoms in the planet's upper atmosphere. To compute the transmissivity along a line of sight (LOS) one needs to know the relation between the observed I and the source intensity I_0. The intensities can be written as a function of frequency, f in units of [s^{-1}] as

$$I(f) = I_0(f)e^{-\tau(f)} \tag{4.4}$$

with the frequency dependent optical depth $\tau(f) = \sigma(f)N$, a frequency depentend cross section $\sigma(f)$ and the atmospheric H column content N in units of m^{-2}. As shown in the right panel of Fig. 4.2, by computing the transmissivity $T = I/I_0$ along the LOS, effects of line broadening in the velocity spectrum of the hydrogen atoms $u(v) = N\hat{u}(v)$ can be studied. It should be noted that N depends on the atmospheric altitude where the H$_2$ molecules are dissociated into H atoms. Thus, future exoplanet atmosphere characterization missions such as EcHO will obtain data, which can be used to locate this transition region in a planet's lower thermosphere. Together with such data, UV transit observations and advanced numerical modeling techniques the upper atmosphere structure and plasma environment around exoplanets can be characterized.

Hydrogen cloud and ENA sensitivity studies for the Earth-like hydrogen-rich exoplanets, which are exposed to high EUV fluxes and denser plasma environments have also been studied recently [49]. These authors showed that similar Lyman-α transit observations as on HD 209458b and HD 189733b may shed some light on the early atmosphere evolution of terrestrial planets, including Venus and Earth.

Because of the large size of the Sun-like G-stars, the Earth-like planets which may be surrounded by hydrogen clouds and ENAs, one can only study the hydrogen exosphere formation around EUV exposed Earth-like planets within active and moderate M-type dwarf stars. In such a case it was found that transit-contrast scenarios of the Earth-type exoplanets with expanded upper atmospheres and resulting hydrogen and ENA-clouds are comparable with that of hot Jupiters in orbits of Sun-type G stars [49].

In the following section, future observations of hydrogen ENA-clouds around Earth-size exoplanets in orbits around M-stars and modeling efforts will be discussed and shown that dwarf stars could be used as a proxy for obtaining data which are important for understanding the evolution of the atmospheres of early Solar System planets.

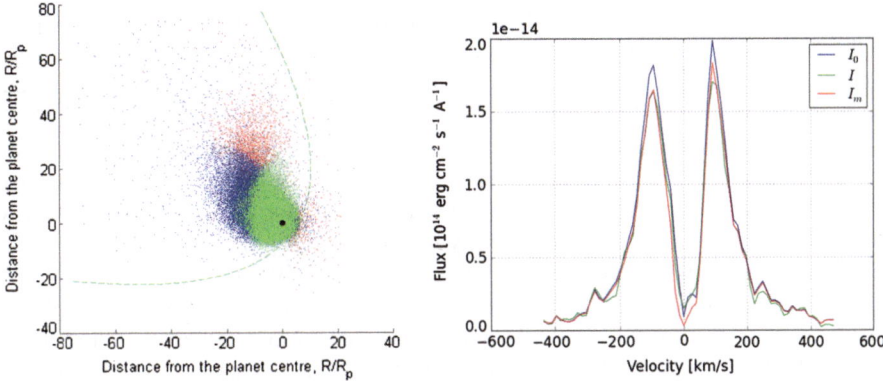

Fig. 4.2 *Left panel*: modeled stellar-wind interaction with HD 209458b with an assumed magnetopause (*dashed green line*) sub-stellar obstacle at about $4.7 R_{pl}$. Shown from above, perpendicular to HD 20948b's orbital plane as seen from the Earth, along the direction of the x-axis. The stellar wind protons are not plotted but flow around the magnetopause obstacle and interact with the planetary hydrogen exosphere (*green dots*) and produce energetic neutral atoms (ENAs: *red and blue dots*). *Right panel*: modeled attenuation spectra and comparison with the HST/STIS observations [28]. Observed profile before transit (*blue line*); observed profile during transit (*green line*). The *red line* is the modeled profile which is computed at the instant of mid-transit by assuming an upper atmosphere number density of 4×10^{13} m^{-3} and an atmospheric temperature of 7,000 K at an altitude of $3 R_{pl}$. The modeled Lyman-α transmissivity spectra includes the effect of line broadening and ENA production. By assuming these particular atmospheric input parameters the model results indicate that broadening alone can explain most of the absorption (courtesy of K. G. Kislyakova)

4.3 Observations of Hydrogen Clouds and Extended Upper Atmospheres Around Terrestrial Exoplanets

The observation and study of EUV heated, extended non-hydrostatic upper atmospheres around terrestrial exoplanets will enhance our knowledge of the stellar radiation and plasma interaction with terrestrial exoplanets as well as the evolution of the Earth-type planetary atmospheres. If one knows the size and mass of an exoplanet by observing the size of a planet's extended upper atmosphere and by determining the velocities of the surrounding hydrogen atoms, conclusions can be drawn with respect to their origin. An observation of a transiting planet with an atomic hydrogen exosphere which extends to $\sim 10\ R_{Earth}$ of an Earth-like planet inside the habitable zone is not possible for G stars [49]. The main reason is that the ratio of planetary to stellar radius requires a very high signal-to-noise ratio (SNR), so that many transits are needed for achieving a very low SNR. Because there is only one transit per year for a planet inside an orbit of a G-star habitable zone, the required SNR cannot be reached with present time observatories.

On the other hand M-type stars are the most likely suitable targets for the observation of extended hydrogen atmospheres around the Earth-size exoplanets because of their small size and related luminosity they provide the unique opportunity to investigate terrestrial planets in the habitable zone, due to the short periods. Due to the large number of M-type stars in our neighborhood and their long periods of higher stellar activity in comparison to G-stars, one can expect that M stars represent the most promising candidates for the detection of hydrogen exospheres and ENA-clouds and the subsequent study of the interaction between the host star and the planets' upper atmosphere. Transit follow-up observations in the UV-range of terrestrial exoplanets around M-type stars with space observatories such as the WSO-UV [50] would provide a unique opportunity to shed more light on the early evolution of Earth-like planets, including those of our own Solar System.

As discussed in previous sections, extended thermospheres and related exospheres will be produced if hydrogen atoms populate the upper atmosphere on a terrestrial planet which is exposed to EUV fluxes ≥ 1 EUV compared to that of the present Sun. Depending on the planet's gravity and the main species in its upper atmosphere as a consequence the upper atmosphere can expand beyond a possible magnetopause and exospheric atoms will be ionized and as a result ENAs are produced via charge exchange collisions between the charged particle flow of the planets host star.

The stellar-wind plasma flow can be described with magnetohydrodynamic or with particle flow models. In terms of modeling, a multi-dimensional plasma flow model that self-consistently includes an ionosphere could be applied to study the interaction of the stellar wind with the planetary environment. However, apart from the modeling and computational complexity, and at the present state of minor knowledge of exoplanetary atmospheric composition and magnetosphere environments it is justified to apply a particle model where one can consider meta-particles of real plasma [20–22].

Outside a planetary obstacle which can be a magnetopause or an ionopause, collisions with an UV photon and charge exchange between a stellar-wind proton as well as with a photoionization or electron impact ionization or elastic collisions with another hydrogen atom on an outward flowing exospheric atom can occur. From the ionization of planetary neutral atoms one can obtain knowledge of the ion loss rate under the studied stellar activity conditions. The stellar-wind flow can be described as a gas dynamic flow around a planetary obstacle [51] where stellar-wind protons which reach the planetary obstacle can then be reflected at the obstacle boundary so that the deflected stellar-wind flow around the obstacle can be modeled. Such models also include external forces such as planetary gravity, Coriolis force, radiation pressure, and gravity of the host star. Photoionization occurs when the planetary neutral atoms are outside the optical shadow behind the planet. The collision processes outside the obstacle can be modeled by a Direct Simulation Monte Carlo (DSMC)-technique where the computational domain can be divided into cells around the planet. After each time step the atoms that belong to the same cell are considered for a hard sphere collisions.

Table 4.1 gives the input parameters for the examples of the modeled hydrogen exospheres shown in Fig. 4.3. The model results in Fig. 4.3 show in the left panel the present Earth geocorona at 1 AU for moderate solar wind conditions and in the right panel an extended hydrogen exosphere around a hydrogen-dominated Earth-size and mass planet within the habitable zone of an M-type dwarf star at ~ 0.24 AU.

Table 4.1 Radius, density, and temperature inner boundary input parameters for the hydrogen exosphere around the Earth in 1 AU and a hydrogen-rich Earth-size and mass planet within the habitable zone of a low mass M star at 0.24 AU which is exposed to a ten times higher EUV flux compared to that of the present solar value

Parameter	Geocorona	H-rich exo-Earth
EUV/EUV_{Sun}	1	10
Star radius (R_{star})	$1 R_{Sun}$	$0.45 R_{Sun}$
Star mass (M_{mass})	$1 M_{Sun}$	$0.45 M_{Sun}$
Planet radius (R_{Earth})	1	1
Planet radius (M_{Earth})	1	1
r_{ib} (R_{Earth})	1.078	10.5
n_{ib} (cm^{-3}) (H atoms)	7×10^4	5×10^4
T_{ib} (K)	900	485
v_{sw} (km s^{-1})	400	330
n_{sw} (cm^{-3})	8	250
T_{sw} (K)	10^6	10^6

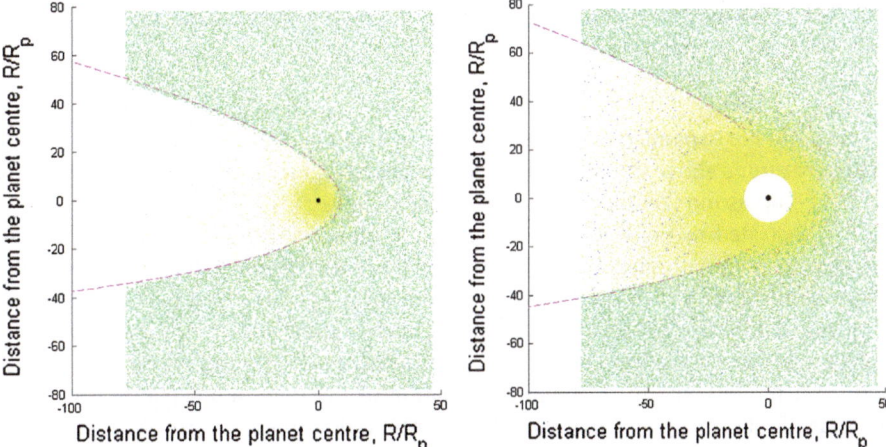

Fig. 4.3 *Left panel*: modeled solar wind flow (protons: *green dots*) around the Earth's magnetosphere (*dashed line*) and interaction with the simulated present day geocorona around the Earth (hydrogen atoms: *yellow dots*). The majority of the Earth's hydrogen exosphere lies within the magnetopause boundary. *Right panel*: modeled stellar wind interaction with an Earth-like hydrogen-rich a planet exposed to 10 EUV within the habitable zone of an M-star (courtesy of K. G. Kislyakova)

One can see from Fig. 4.3 that Earth's geocorona mainly ends within the subsolar magnetopause boundary at about $9 R_{Earth}$ above the surface. In case of the hydrogen-rich exo-Earth which is exposed to a 10 times higher EUV flux, the exobase expands to about $10.5 R_{Earth}$ and the hydrogen exosphere interacts strongly with the stellar wind plasma. In such a case the planet's magnetic moment is most likely too weak to form a magnetosphere which will protect the exosphere. The stellar plasma will

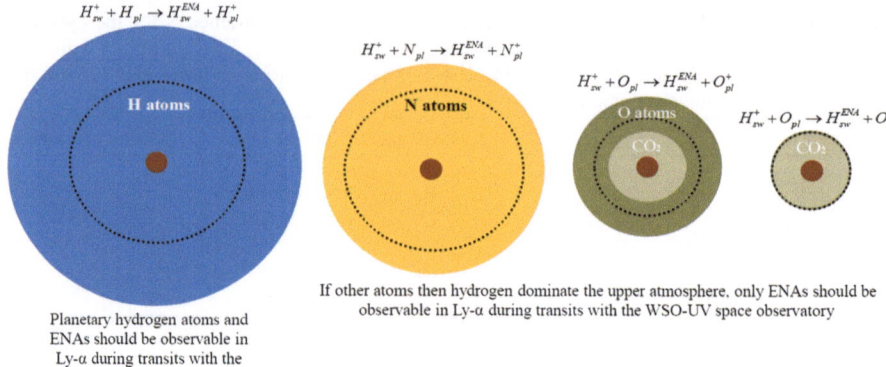

Fig. 4.4 Illustration of EUV-heated and expanded H-, N-, O/CO$_2$- and CO$_2$-dominated upper atmospheres of terrestrial exoplanets and the expected ENA production. The *dotted circles* mark the exobase level. Beyond the exobase, charge exchange between exospheric atoms and stellar wind protons takes place. By knowing the mass and size of the particular planet observations of hydrogen exospheres and/or ENA clouds and the characterization of the corresponding planetary obstacle location should result in an information of the upper atmosphere structure and hence its main thermosphere constituent

flow near to the expanded exobase level where it will be deflected around the planet. Planetary atoms above this obstacle will produce ENAs via charge exchange and can be picked up as ions by the stellar wind plasma flow.

As illustrated in Fig. 4.4, depending on the main species (H, N, O, or CO$_2$) in the thermosphere a higher EUV flux will result in different expansion levels of the upper atmosphere [52–55]. If one knows the mass and size of a particular planet, high resolution UV observations in the Lyman-α emission line in and out of the transit together with advanced numerical exosphere stellar-wind plasma interaction and ENA models one may obtain information of the main atmospheric species in the planet's thermosphere. In case of a hydrogen-dominated upper atmosphere the expected absorption of the host star's Lyman-α emission will contain the slowly outward flowing planetary hydrogen atoms as well as the produced faster moving ENAs around the expanded thermosphere. In case hydrogen does not dominate the upper atmosphere, other species such as N, O, or CO$_2$ will determine the expansion of the planet's upper atmosphere and UV transit observations may show only fast moving hydrogen ENAs which are produced around the planet's obstacle and possible other species such as carbon or oxygen.

M-type dwarf stars with transiting terrestrial planets will be good candidates for such studies, because they represent the majority of stars in our neighborhood, and exhibit strong stellar activity, including in the Lyman-α range for very long time periods. Further, the ratio of the star-to-planet radius is also more favorable towards terrestrial-type planets with extended hydrogen-rich upper atmospheres, making them perfect targets for proxies which will shed more light on how the atmosphere of early terrestrial planets with different sizes and masses survived during the first hundreds of million years of the active young periods of their host stars.

Since the past years the search for the Earth-like exoplanets around low-mass M-type stars has become one of the major activities in stellar astrophysics. Dedicated ground-based telescope instruments are currently planned and developed for the discovery of Earth-mass planets around M dwarfs are for instance the Calar Alto high-Resolution search for M dwarfs with Exoearths with Near-infrared and optical Échelle Spectrographs (CARMENES) [56], and the Habitable Zone Planet Finder (HZPF) [57]. Besides ground-based M star planet search projects such as CARMENES, or HZPF, ESAs planned PLATO mission will also be dedicated to detect transiting terrestrial exoplanets inside habitable zones [58, 59].

The search will include planets of different orbital periods and various sizes, with a focus on about 20,000 bright ($m_V \leq 11$ mag) solar-type host stars. Besides this sample, more then 245,000 cool dwarf and subgiant stars with 11 mag $< m_V < 13$ mag will be monitored to detect terrestrial planets. Moreover the space observatory will also search for exoplanets around more than 10,000 nearby M-type stars with magnitudes down to $m_V = 15$–16 mag.

After the discovery of the Earth-size or super-Earth-size exoplanets from ground-based telescopes or space observatories, follow-up transit observations with the upcoming WSO-UV space telescope could be carried out. The WSO-UV is a 1.7 m telescope devoted to UV spectroscopy and imaging which is included in the Federal Space Program of the Russian Federation with a launch date around 2015–2016 [50, 60] and should be capable to observe Earth-size exoplanets with extended hydrogen exospheres or ENA-clouds if they orbit around small stars [49, 61, 62].

Thus, transit UV follow-up observations of terrestrial exoplanets, which orbit around stars with bright enough Lyman-α emission around M-type stars with space observatories such as the WSO-UV will provide a unique opportunity to shed more light on the early evolution of Earth-like planets, including those of our own Solar System.

References

1. Hovestadt, D., Scholer, M.: Radiation belt-produced energetic hydrogen in interplanetary space. J. Geophys. Res. **81**, 5039–5042 (1976)
2. Krimigis, S.M., Kohl, J.W., Armstrong, T.P.: The magnetospheric contribution to the quiet-time low energy nucleon spectrum in the vicinity of Earth. Geophys. Res. Lett. **2**, 457–460 (1975)
3. Fuselier, S.A., Funsten, H.O., Heirtzler, D., Janzen, P., Kucharek, H., McComas, D.J., Möbius, E., Moore, T.E., Petrinec, S.M., Reisenfeld, D.B., Schwadron, N.A., Trattner, K.J., Wurz, P.: Energetic neutral atoms from the Earth's subsolar magnetopause. Geophys. Res. Lett. **37**(13), (2010)
4. Collier, M.R., Moore, T.E., Ogilvie, K.W., Chornay, D., Keller, J.W., 14 co-authors: Observations of neutral atoms from the solar wind. J. Geophys. Res. **106**, 24893–24906 (2001)
5. Futaana, Y., Barabash, S., Grigoriev, A., Holmström, M., Kallio, E., 42 co-authors: First ENA observations at Mars: ENA emissions from the martian upper atmosphere. Icarus **182**, 424–430 (2006)

6. Gunell, H., Kallio, E., Jarvinen, R., Janhunen, P., Holmström, M., Dennerl, K.: Simulations of solar wind charge exchange X-ray emissions at Venus. Geophys. Res. Lett. **34**, L03107 (2007)
7. Mura, A., Orsini, S., Milillo, A., Kallio, E., Galli, A.: and 30 co-authors: ENA detection in the dayside of Mars: ASPERA-3 NPD statistical study. Planet. Space Sci. **56**, 840–845 (2008)
8. Galli, A., Wurz, P., Bochsler, P., Barabash, S., Grigoriev, A., Futaana, Y., Holmström, M., 44 co-authors: First observation of energetic neutral atoms in the Venus environment. Planet. Space Science **56**, 807–811 (2008)
9. Lichtenegger, H.I.M., Lammer, H., Kulikov, Yu.N., Kazeminejad, S., Molina-Cuberos, G.H., Rodrigo, R., Kazeminejad, B., Kirchengast, G.: Effects of low energetic neutral atoms on martian and venusian dayside exospheric temperature estimations. Space Sci. Rev. **125**, 469–501 (2006)
10. Barth, C.A., Fastie, W.G., Hord, C.W., Pearce, J.B., Kelly, K.K., Stewart, A.I., Thomas, G.E., Anderson, G.P., Raper, O.F.: Mariner 6: ultraviolet spectrum of Mars upper atmosphere. Science **165**, 1004–1005 (1969)
11. Barth, C.A.: Interpretation of the Mariner 5 Lyman alpha measurements. J. Atmos. Sci. **25**, 564–567 (1968)
12. Keating, G.M., Bougher, S.W., Zurek, R.W., Tolson, R.H., Cancro, G.J., 23 co-authors: The structure of the upper atmosphere of Mars: in situ accelerometer measurements from Mars Global Surveyor. Science **279**, 1672–1676 (1998)
13. Lichtenegger, H.I.M., Lammer, H., Vogl, D., Bauer, S.J.: Possible temperature effects of energetic neutral hydrogen atoms on the martian exosphere. Adv. Space Res. **33**, 140–144 (2004)
14. Mitchell, D.G., Paranicas, C.P., Mauk, B.H., Roelof, E.C., Krimigis, S.M.: Energetic neutral atoms from Jupiter measured with the Cassini magnetospheric imaging instrument: time dependence and composition. J. Geophys. Res. **109**(A9), A09S11 (2004)
15. Dandouras, J., Amsif, A.: Production and imaging of energetic neutral atoms from Titan's exosphere: a 3-D model. Planet. Space Sci. **47**, 1355–1369 (2004)
16. Mitchell, D.G., Brandt, P.C., Roelof, E.C., Dandouras, J., Krimigis, S.M., Mauk, B.H.: Energetic neutral atom emissions from Titan interaction with Saturn's magnetosphere. Science **308**, 989–992 (2005)
17. Gruntman, M., Roelof, E.C., Mitchell, D.G., Fahr, H.J., Funsten, H.O., McComas, D.J.: Energetic neutral atom imaging of the heliospheric boundary region. J. Geophys. Res. **106**, 15767–15782 (2001)
18. Wood, B.E., Müller, H.-R., Zank, G., Linsky, J.L.: Measured mass loss rates of solar-like stars as a function of age and activity. ApJ **574**, 412–425 (2002)
19. Wood, B.E., Müller, H.-R., Zank, G.P., Linsky, J.L., Redfield, S.: New mass loss measurements from astrospheric Ly-α absorption. ApJ **628**, L143–L146 (2005)
20. Holmström, M., Ekenbäck, A., Selsis, F., Penz, T., Lammer, H., Wurz, P.: Energetic neutral atoms as the explanation for the high-velocity hydrogen around HD 209458b. Nature **451**, 670–679 (2010)
21. Ekenbäck, A., Holmström, M., Wurz, P., Grießmeier, J.-M., Lammer, H., Selsis, F., Penz, T.: Energetic neutral atoms around HD 209458b: estimations of magnetospheric properties. ApJ **709**, 670–679 (2010)
22. Lammer, H., Kislyakova, K.G., Holmström, H., Khodachenko, M.L.: Griemeier, J.-M.: Hydrogen ENA-cloud observation and modeling as a tool to study star-exoplanet interaction. Astrophys. Space Sci. **335**, 9–23 (2011)
23. Charbonneau, D., Brown, T.M., Latham, Mayor, D.W.: Detection of planetary transits across a Sun-like star. ApJ 529, L45–L48 (2000)
24. Guillot, T., Burrows, A., Hubbard, W.B., Lunine, J.I., Saumon, D.: Giant planets at small orbital distances. ApJ **459**, L35–L38 (1996)
25. Hubbard, W.B., Burrows, A., Lunine, J.I.: Theory of giant planets. Ann. Rev. Astron. Astrophys. **40**, 103–136 (2002)
26. Lammer, H., Selsis, F., Ribas, I., Guinan, E.F., Bauer, S.J., Weiss, W.W.: Atmospheric loss of exoplanets resulting from stellar X-Ray and extreme-ultraviolet heating. ApJ **598**, L121–L124 (2003)

27. Vidal-Madjar, A.: Lecavelier des Etangs, A., Désert, J.M., Ballester, G.E., Ferlet, R., Hébrard, G., Mayor, M.: An extended upper atmosphere around the extrasolar planet HD209458 b. Nature **422**, 143–146 (2003)
28. Ben-Jaffel, L.: Exoplanet HD 209458b: inated hydrogen atmosphere but no sign of evaporation. ApJ **671**, L61–L64 (2007)
29. Ben-Jaffel, L., Sona Hosseini, S.: On the existence of energetic atoms in the upper atmosphere of exoplanet HD 209458b. ApJ **709**, 1284–1296 (2010)
30. Lecavelier des Etangs, A., Ehrenreich, D., Vidal-Madjar, A., Ballester, G.E., Désert, J.-M., Ferlet, R., Hébrard, G., Sing, D.K., Tchakoumegni, K.-O., Udry, S., 2010. Evaporation of the planet HD 189733b observed in H I Lyman-α. A&A **514**, A72 (2010)
31. Guo, J.H.: Escaping particle fluxes in the atmospheres of close-in exoplanets. I. Model of hydrogen. ApJ **733**, 98, 10 (2011)
32. Penz, T., Erkaev, N.V., Kulikov, Yu.N., Langmayr, D., Lammer, H., Micela, G., Cecchi-Pestellini, C., Biernat, H.K., Selsis, F., Barge, P., Deleuil, M., Léger, A.: Mass loss from "Hot Jupiters"- Implications for CoRoT discoveries, Part II: Long time thermal atmospheric evaporation modeling. Planet. Space Sci. **56**, 1260–1272 (2008)
33. Yelle, R.V.: Aeronomy of extra-solar giant planets at small orbital distances. Icarus **170**, 167–179 (2004)
34. Tian, F., Toon, O.B., Pavlov, A.A., De Sterck, H.: Transonic hydrodynamic escape of hydrogen from extrasolar planetary atmospheres. ApJ **621**, 1049–1060 (2005)
35. García Muñoz, A.: Physical and chemical aeronomy of HD 209458b. Planet. Space Sci. **55**, 1426–1455 (2007)
36. Koskinen, T.T., Yelle, R.V., Lavvas, P., Lewis, N.K.: Characterizing the thermosphere of HD209458 b with UV tranist observations. ApJ **723**, 116–128 (2010)
37. Erkaev, N.V., Kulikov, Yu.N., Lammer, H., Selsis, F., Langmayr, D., Jaritz, G.F., Biernat, H.K.: Roche lobe effects on the atmospheric loss of "Hot Jupiters". A&A **472**, 329–334 (2007)
38. Vidal-Madjar, A., Désert, J., Lecavelier des Etangs, A., Hébrard, G., Ballester, G.E., Ehrenreich, D., Ferlet, R., McConnell, J.C., Mayor, M., Parkinson, C.D.: Detection of oxygen and carbon in the hydrodynamically escaping atmosphere of the extrasolar planet HD 209458b. ApJ **604**, L69–L72 (2004)
39. Linsky, J.L., Yang, H., France, K., Froning, C.S., Green, J.C., Stocke, J.T., Osterman, S.N.: Observations of mass loss from the transiting exoplanet HD 209458b. ApJ **717**, 1291–1299 (2010)
40. Fossati, L., Haswell, C.A., Froning, C.S., Hebb, L., Holmes, S., Kolb, U., Helling, C., Carter, A., Wheatley, P., Cameron, A.C., Loeillet, B., Pollacco, D., Street, R., Stempels, H.C., Simpson, E., Udry, S., Joshi, Y.C., West, R.G., Skillen, I., Wilson, D.: Metals in the exosphere of the highly irradiated planet WASP-12b. ApJ **714**, L222–L227 (2010)
41. Lammer, H., Odert, P., Leitzinger, M., Khodachenko, M.L., Panchenko, M., Kulikov, Yu.N, Zhang, T.L., Lichtenegger, H.I.M., Erkaev, N.V., Wuchterl, G., Micela, G., Penz, A., Biernat, H.K., Weingrill, J., Steller, M., Ottacher, H., Hasiba, J., Hanslmeier, A.: Determining the mass loss limit for close-in exoplanets: what can we learn from transit observations? A&A **506**, 399–410 (2009)
42. Erkaev, N.V., Penz, T., Lammer, H., Lichtenegger, H.I.M., Wurz, P., Biernat, H.K., Griessmeier, J.-M., Weiss, W.W.: Plasma and magnetic field parameters in the vicinity of short periodic giant exoplanets. ApJS **157**, 396–401 (2005)
43. Murray-Clay, R.A., Chiang, E.I., Murray, N.: Atmospheric escape from hot Jupiters. ApJ **693**, 23–42 (2009)
44. Shematovich, V.I.: Suprathermal hydrogen produced by the dissociation of molecular hydrogen in the extended atmosphere of exoplanet HD 209458b. Sol. Syst. Res. **44**, 96–103 (2010)
45. Khodachenko, M.L., Lammer, H., Lichtenegger, H.I.M., Langmayr, D., Erkaev, N.V., Grießmeier, J.M., Leitner, M., Penz, T., Biernat, H.K., Motschmann, U., Rucker, H.O.: Mass loss of "Hot Jupiters": implications for CoRoT discoveries. Part I: the importance of magnetospheric protection of a planet against ion loss caused by coronal mass ejections. Planet. Space Sci. **55**, 631–642 (2007)

46. Li, S-L.: Miller, N., Lin, D.N.C., Fortney, J.J.: WASP-12b as a prolate, inflated and disrupting planet from tidal dissipation. Nature **463**, 1054–1056 (2010)
47. Lammer, H., Kislyakova, K.G., Odert, P., Leitzinger, M., Khodachenko, M.L., Holmström, M., Hanslmeier, A.: Exoplanet upper atmosphere envrionment characterization. In: Richards, M., Hubeny, I. (eds.) From Interacting Binaries to Exoplanets: Essential Modeling Tools, vol. 282, pp. 525–533. Cambridge University Press, Cambridge, Proceedings of the IAU (2012)
48. Lecavelier des Etangs, A., Vidal-Madjar, A., Désert, J.-M.: The origin of hydrogen around HD 209458b. Nature **456**, E1 (2008)
49. Lammer, H., Eybl, V., Kislyakova, K.G., Weingrill, J., Holmström, M., Khodchenko, M.L., Kulikov, Yu.N, Reiners, A., Leitzinger, M., Odert, P., Xian Grüß, M., Dorner, B., Güdel, M., Hanslmeier, A.: UV transit observations of EUV-heated expanded thermospheres of Earth-like exoplanets around M-stars: testing atmosphere evolution scenarios. Astrophys. Space Sci. **335**, 39–50 (2011)
50. Shustov, B., Sachov, M., Gomez de Castro, A.I., Ana, I., Pagano, I.: WSO-UV ultraviolet mission for the next decade. Astrophys. Space Sci. **320**, 187–190 (2009)
51. Spreiter, J.R., Stahara, S.S.: A new predicative model for determining solar wind-terrestrial planet interactions. J. Geophys. Res. **85**, 6769–6777 (1980)
52. Tian, F., Kasting, J.F., Liu, H., Roble, R.G.: Hydrodynamic planetary thermosphere model: 1. The response of the Earth's thermosphere to extreme solar EUV conditions and the significance of adiabatic cooling. J. Geophys. Res. **113**, (2008). doi:10.1029/2007JE002946
53. Tian, F., Solomon, S.C., Qian, L., Lei, J., Roble, R.G.: Hydrodynamic planetary thermosphere model: 2. Coupling of an electron transport/energy deposition model. J. Reophys. Res. **113**, E07005 (2008)
54. Tian, F.: Thermal escape from super Earth atmospheres in the habitable zones of M Stars. ApJ **703**, 905–909 (2009)
55. Lichtenegger, H.I.M., Lammer, H., Grießmeier, J.-M., Kulikov, Yu.N., von Paris, P., Hausleitner, W., Krauss, S., Rauer, H.: Aeronomical evidence for higher CO_2 levels during Earth's Hadean epoch. Icarus **210**, 1–7 (2010)
56. Quirrenbach, A., Amado, P.J., Mandel, H., Caballero, J.A., Ribas, I., Reiners, A., Mundt, R.: and the CARMENES Consortium: CARMENES: Calar Alto high-resolution search for M dwarfs with exo-earths with a near-infrared echelle spectrograph. Astron. Soc. Pac. Conf. Ser. **77356**, 37 (2010)
57. Mahadevan, S., Ramsey, L., Redman, S., Zonak, S., Wright, J., Wolszczan, A., Endl, M., Zhao, B.: The habitable zone planet finder project: a proposed high resolution NIR spectrograph for the Hobby Eberly Telescope (HET) to discover low mass exoplanets around M stars. Astron. Soc. Pac. Conf. Ser. **77356**, 10 (2010)
58. Catala, C., The PLATO team: PLATO: PLAnetary Transits and Oscillations of stars. Exp. Astron. **23**, 329–256 (2009)
59. Catala, C., The PLATO team: PLATO: PLAnetary transits and oscillations of stars. Community Asteros. **158**, 330–336 (2009)
60. Lammer, H., Hanslmeier, A., Schneider, J., Stateva, I.K., 32 co-authors: Exoplanet status report: observation, characterization and evolution of habitable exoplanets and their host stars. Sol. Syst. Res. **44**, 314–335 (2010)
61. Lammer, H., Kislyakova, K.G., Odert, P., Leitzinger, M., Schwarz, R., Pilat-Lohinger, E., Kulikov, Yu.N, Khodachenko, M.L., Güdcl, M., Hanslmcier, A.: Pathways to Earth-likc atmospheres: extreme ultraviolet (EUV)-powered escape of hydrogen-rich protoatmospheres. Orig. Life Evol. Biosph. **41**, 503–522 (2012)
62. Lammer, H., Güdel, M., Kulikov, Yu.N, Ribas, I., Zaqarashvili, T.V., Khodachenko, M.L., Kislyakova, K.G., Gröller, H., Odert, P., Leitzinger, M., Fichtinger, B., Krauss, S., Hausleitner, W., Holmström, M., Sanz-Forcada, J., Lichtenegger, H.I.M., Hanslmeier, A., Shematovich, V.I., Bisikalo, D., Rauer, H., Fridlund, M.: Variability of solar/stellar activity and magnetic field and its influence on planetary atmosphere evolution. Earth Planets Space **63**, 179–199 (2012)

Chapter 5
Conclusion

The atmosphere evolution of terrestrial planets from the origin of protoatmospheres throughout the most active period of a planet's host star until the star's activity decreased to moderate and quiet levels is discussed. The first protoatmosphere will be captured and accumulated hydrogen- and helium-rich gas envelopes from the nebula. Depending on the planetary formation time, the nebula dissipation time, the numbers of additional planets including gas giants in the system, the protoplanet's gravity, its orbit location, and the host star's radiation and plasma environment terrestrial planets may capture tens or even several hundreds of the Earth ocean equivalent amounts of hydrogen around its rocky core.

The second protoatmosphere depends on the initial volatile content of the protoplanet when accretion finished. During the magma ocean solidification depending on the initial water and volatile inventory steam atmospheres with surface pressures ranging from $\sim$100 to several 10^4 bar can be catastrophically outgassed. Finally, secondary atmospheres will be produced by tectonic activity such as volcanos and by the delivery of volatiles via large impacts. The origin and initial state of a planet's protoatmosphere, therefore, determines a planet's atmospheric evolution and finally if the planet will evolve to an Earth-analog class I habitat or not. Thus, one can conclude that the current atmospheres of Venus, the Earth and Mars have resulted from the action of

- possible accumulation of nebular gases (H, H_2, He) during protoplanetary growth,
- catastrophic release of volatiles and atmosphere formation after the magma ocean solidification,
- EUV ($\lambda \approx$ 2–120 nm) and plasma driven thermal and non-thermal atmospheric escape processes of volatiles,
- impact erosion and delivery,
- impact delivery and mantle outgassing throughout later evolutionary epochs, as well as
- inorganically or organically enhanced carbonate formation due to CO_2 weathering processes.

H. Lammer, *Origin and Evolution of Planetary Atmospheres*,
SpringerBriefs in Astronomy, DOI: 10.1007/978-3-642-32087-3_5,
© The Author(s) 2013

To understand how a planetary atmosphere evolved during its lifetime one has also to understand how the radiation environment and the stellar wind of its host star evolved with time. The stellar wind and the host star's X-ray, SXR, and EUV radiation act as a permanent nonlinear force at planetary upper atmospheres. Observations of radiation in a wavelength range from X-rays to far-UV by XMM-Newton, ASCA, ROSAT, EUVE, FUSE, HST, IUE, and UBVRI satellites of Sun-like stars with different age deliver evidence that the young Sun was rotating more than 10 times than its present rate and had correspondingly strong dynamo-driven high-energy emissions which resulted in EUV emissions, up to several 100 times stronger than that of the present Sun. Furthermore, empirical correlations of stellar mass loss rates with X-ray surface flux values indicate that the solar/stellar wind may have been ≥ 100 times more dense at the Earth's orbital distance.

The efficiency of the solar/stellar forcing is essentially inversely proportional to the square of the distance to the planet's host star. From that, it follows that the closer a planet orbits around its host star, the more efficient are the atmospheric escape processes. The main effects caused by the stellar radiation and plasma environment on the atmospheres of an effected planet are to ionize, chemically modify, heated, expand, and slowly erode the upper atmosphere throughout the lifetime of a planet. The highest thermal and non-thermal atmospheric escape rates are obtained during the early active phase of the planet's host star, which lasts about 500 Myr for G-type stars and could last up to a few Gyr at M-type dwarf stars.

Besides the orbital location, a planet's gravity constitutes an additional major protection mechanism especially for thermal escape of its atmosphere, while the non-thermal escape processes are affected on a weaker scale. Ionospheric plasma energization and ion pick up represent two categories of non-thermal escape processes that may bring planetary ions beyond escape velocity. For present time, these energization processes have been studied around Venus, the Earth and Mars by various plasma instruments on various spacecraft since the past decades.

At the present Earth, with its strong intrinsic magnetic moment provides a kind of "magnetic protection" against ion pick up. Because the solar wind plasma flow is deflected around the magnetopause at a distance of about $9-10 R_{Earth}$. In the Earth's past when the upper atmosphere expanded most likely above the magnetopause due to strong EUV heating the magnetosphere did not protect the exosphere against strong solar wind erosion. During that early period of the planet, the solar wind interaction with its upper atmosphere was comparable to that of present Mars and Venus where the solar wind has more direct access to their topside atmospheres.

However, there are observational and theoretical indications that many planets within the size and mass range from the Earth-type to the so-called super-Earths may not completely get rid of their initial protoatmospheres. One can expect that there may be many super-Earth-type planets which accumulated a dense abiotic oxygen-rich upper atmosphere or remain as hydrogen-rich subNeptune-type bodies. On the other hand, if terrestrial planets lose their dense protoatmospheres or high levels of CO_2 too fast during the high EUV phase of their young host stars, nitrogen-dominated atmosphere may escape to space. The results discussed in this monograph indicate that the Earth's early atmospheric nitrogen inventory should have been protected against high thermal and non-thermal escape rates by higher amounts of IR-cooling CO_2 contents in the thermosphere and/or a dense hydrogen envelopes which remained from the initial protoatmosphere during $\sim$500 Myr after the planet's origin.

The main conclusions drawn on the basis of the stellar evolution and related time dependence of the radiation and plasma energization to atmospheric escape processes can be summarized to

- solar radiation and plasma forcing is efficient in removing volatiles, primarily hydrogen and water and during the most active period of a planet's host star also heavier atmospheric species such as O, C, and N,
- planets orbiting close to the young Sun or star are subject to a strong loss of water, the effect being most profound for Venus,
- depending on the orbital location, small and low mass planets such as Mars can lose their protoatmospheres easier so that a secondary atmospheres produced by tectonic activity or later impacts may become more relevant after the solar/stellar activity decreased to values that the escape flux is lower compared to the outgassed flux,
- depending on the solar/stellar activity and atmospheric main species an intrinsic planetary magnetic field, like the Earth's dipole field provides a shield against ion pick up loss, while ion outflow over the planet's polar areas can also be high,
- on many planets the efficiency of atmospheric escape processes are most likely too weak to remove dense hydrogen-rich protoatmospheres over their lifetime,
- many planets may have upper atmospheres which are populated by abiotic oxygen atoms related to dissociated H_2O molecules,
- terrestrial exoplanets which represent the Earth analogue class I habitats are definitely more rare compared to planets with dense $H/O/H_2O$ and CO_2-type atmospheres, or planets which are covered by deep water oceans.

Future exoplanet atmosphere characterization projects and UV transit observations of terrestrial exoplanets within orbits of M-type dwarf stars, together with advanced numerical modeling techniques will help to refute or confirm the atmosphere evolution hypotheses which have been addressed in this brief monograph during the not too distant future.

Literature

1. Atreya, S.K., Pollack, C.K., Matthews, M.S. (eds.): Origin and Evolution of Planetary and Satellite Atmospheres. The University of Arizona Press, Tucson (1989)
2. Bauer, S.J., Lammer, H.: Planetary Aeronomy—Atmosphere Environments in Planetary Systems. Springer, New York (2004)
3. Barstow, M.A., Holberg, J.B.: Extreme Ultraviolet Astronomy. Cambridge University Press, Cambridge (2003)
4. Beaulieu, J.-P., Dieters, S., Tinetti, G.: Molecules in the atmospheres of extrasolar planets. Astronomical Society of the Pacific Conference Series, vol. 450, (2011)
5. Chamberlain, J.W., Hunten, D.M.: Theory of Planetary Atmospheres, 2nd edn. Academic Press, San Diego (1987)
6. Coudé du Foresto, V., Gelino, D.M., Ribas, I.: Pathways towards habitable planets. Astronomical Society of the Pacific Conference Series, vol. 430, (2010)
7. Dvorak, R.: Extrasolar Planets—Formation, Detection and Dynamics. Physics Textbook. Wiley-VCH, Weinheim (2008)
8. Fridlund, M., Lammer, H.: Astrobiology—habitability primer. Astrobiology **10**(1), 1–4 (2010)
9. Güdel, M.: The sun in time: activity and environment. Living Rev. Solar Phys. vol. 4(3). http://solarphysics.livingreviews.org/Articles/lrsp-2007-3/, (2007)
10. Hanslmeier, A.: Water in the Universe. Springer, New York (2011)
11. Horneck, G., Baumstark-Khan, C.: Astrobiology: The Quest for the Conditions of Life. Springer, New York (2002)
12. Kasting, J.F.: How to Find a Habitable Planet. Princeton University Press, Princeton (2009)
13. Kasting, J.F., Catling, D.: Evolution of a habitable planet. Ann. Rev. Astron. Astrophys. **41**, 429–463 (2003). doi:10.1146/annurev.astro.41.071601.170049
14. Kivelson, M.G., Russel, C.T.: Introduction to Space Physics. Cambridge University Press, Cambridge (1995)
15. Lammer, H., Kasting, J.F., Chassefière, E., Johnson, R.E., Kulikov, Yu.N., Tian, F.: Atmospheric escape and evolution of terrestrial planets and satellites. Space Sci. Rev. **139**, 399–436 (2008)
16. Lammer, H., Bredehöft, J.H., Coustenis, A., Khodachenko, M.L., Kaltenegger, L., Grasset, O., Prieur, D., Raulin, F., Ehrenfreund, P., Yamauchi, M., Wahlund, J.-E., Grießmeier, J.-M., Stangl, G., Cockell, C.S., Kulikov, Yu.N., Grenfell, L., Rauer, H.: What makes a planet habitable? Astron. Astrophys. Rev. **17**, 181–249 (2009)
17. Lewis, J.S., Prinn, R.G.: Planets and Their Atmospheres. Academic Press (1989)

H. Lammer, *Origin and Evolution of Planetary Atmospheres*,
SpringerBriefs in Astronomy, DOI: 10.1007/978-3-642-32087-3,
© The Author(s) 2013

18. Lunine, I.J.: Earth: Evolution of a Habitable World. Cambridge University Press, Cambridge (1999)
19. Montesinos, B., Giménez, Á., Guinan, E.F.: The evolving Sun and its influence on planetary environments. Astronomical Society of the Pacific Conference Series, vol. 269, (2002)
20. Richards, M.T., Hubeny, I.: From interacting binaries to exoplanets: essential modeling tools. In: Proceedings of the International Astronomical Union Symposium No. 282, 18–22 July 2011, Cambridge University Press, Tatranská Lomnica, Slovakia (2012)
21. Schunk, R.W., Nagy, A.: Ionospheres: Physics, Plasma Physics and Chemistry. Cambridge University Press, Cambridge (2002)
22. Seager, S.: Exoplanet Atmospheres—Physical Processes. Princeton University Press, Princeton (2010)
23. Ward, P.D., Brownlee, D.: Rare Earth—Why Complex Life is Uncommon in the Universe. Copernicus, New York (2000)

Index